Artificial Intelligence in Textile Engineering

THE TEXTILE INSTITUTE BOOK SERIES

Incorporated by Royal Charter in 1925, The Textile Institute was established as the professional body for the textile industry to provide support to businesses, practitioners, and academics involved with textiles and to provide routes to professional qualifications through which Institute Members can demonstrate their professional competence. The Institute's aim is to encourage learning, recognise achievement, reward excellence, and disseminate information about the textiles, clothing and footwear industries and the associated science, design and technology. It has a global reach, with individual and corporate members in over 80 countries.

The Textile Institute Book Series supersedes the former "Woodhead Publishing Series in Textiles" and represents a collaboration between The Textile Institute and Elsevier aimed at ensuring that Institute Members and the textile industry continue to have access to high calibre titles on textile science and technology.

Books published in The Textile Institute Book Series are offered on the Elsevier website at: store.elsevier.com and are available to Textile Institute Members at a discount. Textile Institute books still in print are also available directly from the Institute's website at: www.textileinstitute.org

To place an order, or if you are interested in writing a book for this series, please contact Sophie Harrison, Acquisitions Editor: s.harrison2@elsevier.com

Recently Published and Upcoming Titles in the Textile Institute Book Series:

Smart Textiles from Natural Resources, 1st Edition, Md. Ibrahim H. Mondal, 978-0-44-315471-3

Advances in Plasma Treatment of Textile Surfaces, 1st Edition, Shahid Ul Islam, Aminoddin Haji, 978-0-44-319079-7

The Wool Handbook: Morphology, Structure, Property and Application, 1st Edition, Seiko Jose, Sabu Thomas, Gautam Basu, 978-0-32-399598-6

Natural Dyes for Sustainable Textiles, 1st Edition, Padma Shree Vankar, Dhara Shukla, 978-0-32-385257-9

Digital Textile Printing: Science, Technology and Markets, 1st Edition, Hua Wang, Hafeezullah Memon, 978-0-44-315414-0

Textile Calculation: Fibre to Finished Garment, 1st Edition, R. Chattopadhyay, Sujit Kumar Sinha, Madan Lal Regar, 978-0-32-399041-7

Advances in Healthcare and Protective Textiles, 1st Edition, Shahid Ul Islam, Abhijit Majumdar, Bhupendra Singh Butola, 978-0-32-391188-7

Fabrication and Functionalization of Advanced Tubular Nanofibers and their Applications, 1st Edition, Baoliang Zhang, Mudasir Ahmad, 978-0-32-399039-4

Functional and Technical Textiles, 1st Edition, Subhankar Maity, Kunal Singha, Pintu Pandit, 978-0-32-391593-9

The Textile Institute Book Series

Artificial Intelligence in Textile Engineering

Basic Concepts and Applications

Anindya Ghosh
Government College of Engineering and Textile Technology,
Berhampore, West Bengal, India

Subhasis Das
Department of Textile Technology, Government College of Engineering
and Textile Technology, Berhampore, West Bengal, India

Bapi Saha
Department of Mathematics, Government College of Engineering
and Textile Technology, Berhampore, West Bengal, India

WP
WOODHEAD
PUBLISHING
ELSEVIER   An imprint of Elsevier

Woodhead Publishing is an imprint of Elsevier
50 Hampshire Street, 5th Floor, Cambridge, MA 02139, United States
125 London Wall, London EC2Y 5AS, United Kingdom

ISBN: 978-0-443-15395-2 (print)
ISBN: 978-0-443-22124-8 (online)

For information on all Woodhead publications
visit our website at https://www.elsevier.com/books-and-journals

Publisher: Matthew Deans
Acquisitions Editor: Sophie Harrison
Editorial Project Manager: Tessa Kathryn
Production Project Manager: Maria Bernard
Cover Designer: Vicky Pearson Esser

Typeset by STRAIVE, India

Working together
to grow libraries in
developing countries

www.elsevier.com • www.bookaid.org

Contents

6. Nature-inspired optimization algorithms

7. Hybrid artificial intelligence systems

Preface

In the annals of human civilization, three milestone events stand out in seminal significance: renaissance, industrial revolution, and the latest being artificial intelligence (AI). A few extraordinary visionaries of the 19th century such as Jule Verne and H.G. Wells indulged in the wildest flight of fancy, and their narrative mastery often had the tantalizing effect of turning the imponderable into likelihood or even reality. And truly at the dawn of the 21st century we are poised to witness the transcendence of fancy into fact on the sheer strength of human ingenuity and intelligence. In such a grand scheme of transformation, the prime mover is AI, a synergic combination of computer systems and algorithms simulated to imitate human intelligence. AI has boundless potential and every walk of human life or all that are mundane, terrestrial or extraterrestrial stand to reap benefit of it. AI is the indispensable component of the ongoing concept of Industry 4.0 representing the paradigm shift in manufacturing and production. When every existing science and technology gears up to integrate AI in its repertoire, the textile world must be on the bandwagon.

In the realm of textile engineering, where precision, efficiency, and innovation converge, the acquisition of AI has betokened a transformative force. This book, *Artificial Intelligence in Textile Engineering: Basic Concepts and Applications* attempts to demystify the general concept of AI with its various domains such as artificial neural networks (ANN), support vector machines (SVM), fuzzy logic, rough set, genetic algorithm (GA), ant colony optimization (ACO), particle swarm optimization (PSO), and hybrid intelligence systems. In the introductory chapter, readers are led through a lucid yet substantial discourse before they are exposed to the rigors of individual domains. Chapter 2 sheds light on the fundamental concepts of ANNs along with their architecture and training methods. Chapter 3 presents basic principles of support vector classification and support vector regression. Chapter 4 focuses on the subject of fuzzy logic, a potent tool dealing with uncertainty and vagueness in the data by assigning degrees of membership to elements. Chapter 5 addresses the concept of rough set theory representing uncertainty in data by dividing it into two subsets—lower approximation and upper approximation. Chapter 6 is devoted to some of the finest optimization techniques based on computational methods seeking inspiration from nature. The basic concepts of GA emulating the principles of evolution, PSO inspired by the social behavior of bird flocking, and ACO taking after the ants' capability of finding the shortest route from the nest to a food source have been discussed in appropriate detail. Chapter 7 explains some of the hybrid AI systems. The adaptive neuro-fuzzy inference system and genetic fuzzy expert system are brought into consideration because of their usefulness.

Authors have gleaned a large body of real-world problems from the diverse areas of textiles and catered to them as solved ones for the readers. Delving into the practical aspects, the book elucidates how AI is revolutionizing various facets of textile engineering. Readers will encounter case studies and examples that showcase the application of AI in quality control, production optimization, and other critical areas within the textile industry.

Authors quite ambitiously stepped out of their way to make this book useful for learners and practitioners in other science and technological domains as well.

For each of the technique, MATLAB® coding has been provided, which will surely ease out students' practical difficulties during problem-solving. These MATLAB coding will definitely motivate the learners to devote themselves to applications pertaining to AI.

As the authors of this book, our aspiration is to provide readers with a comprehensive guide that not only demystifies the complexities of AI but also inspires a vision for its transformative potential within the realm of textile engineering. This book will serve as a valuable resource for students, researchers, and professionals in the field of AI as they navigate their journey of exploration and innovation.

Anindya Ghosh
Subhasis Das
Bapi Saha

Chapter 1

Introduction

1.1 Introduction

The invention of the von Neumann digital computer initiated an ever-growing attempt—how to program them to learn to improve automatically with experience like human beings. The impact has been proven to be dramatic with the successful endeavor of developing an intelligent machine that embodies the high processing power of a computer and sensible decision making of humans through experience. An intelligence machine can modify its processing on the basis of newly acquired information and thereafter make decisions in a rightfully sensible manner when presented with inputs. The McCulloch-Pitts neuron model in 1943 laid the foundation for subsequent developments in the field of "artificial intelligence (AI)." The term AI was first coined by John McCarthy in 1956 in the Dartmouth Workshop, which is often considered the birth of AI as a field. AI may be defined as the ability of machines to perform tasks that are typically associated with human intelligence, such as learning and problem-solving. Oxford Dictionary defines it as the study and development of computer systems that can copy intelligent human behavior. Machine learning is a subfield of AI concerned with the construction of programs that learn from experience and make predictions or classifications.

It is an established fact that most of the real-world problems are too complex to model mathematically and there exist many physical tasks that cannot be solved by classical programming techniques. The machine learning is gaining strategic importance by solving such types of problems where precision is considered to be secondary and we are interested primarily in acceptable solutions. By and large, machine learning is a collection of many methodologies, such as artificial neural network, support vector machine (SVM), fuzzy logic, rough set, genetic algorithm, particle swarm optimization, ant colony optimization, and their different hybrid forms. Different members of this family are able to perform various types of tasks. For example, artificial neural network and SVM are the potential tools for prediction and classification; fuzzy logic and rough set are the powerful tools for dealing with imprecision and uncertainty; genetic algorithm, particle swarm optimization, ant colony optimization, etc., are the important tools for search-based optimization. In combined techniques, either two or more tools are amalgamated to get the advantage of both.

Artificial Intelligence in Textile Engineering. https://doi.org/10.1016/B978-0-443-15395-2.00007-7

AI applications have become increasingly prevalent in everyday life, from healthcare to finance and manufacturing industries. With the advent of very high computational speed, the machine learning has empowered scientists and technologists from diverse engineering disciplines. Focusing attention on the sphere of textile engineering, the rapidly bourgeoning sway of machine learning is now well recognized in the selection of raw material, setting of process parameters, classification of patterns, prediction of the properties of various fibrous products, and their design optimization. The complex raw material-fibrous product relationship invokes artificial neural network, support vector regression, and different hybrid forms as the potent methods. Fuzzy logic is an appealing tool for textile product engineering as it can able to handle imprecision that is present in the textile data. For example, a spinner often uses terms such as "fine" and "coarse" to assess the fiber and yarn count, although these terms do not constitute a well-defined boundary. Optimization algorithms such as genetic algorithm, particle swarm optimization, and ant colony optimization seem to be the right approaches to designing the parameters of a textile product for the purpose of achieving minimum manufacturing cost or maximum strength. Support vector classifier is the foremost tool to recognize the variant textile patterns.

Supervised learning is a fundamental paradigm in machine learning where an algorithm learns from a dataset. In this approach, the algorithm is provided with input-output pairs, where the inputs represent the data, and the outputs are the corresponding target labels or values. The primary objective of supervised learning is to generalize from this training data to make accurate predictions or classifications on new, unseen data.

Prediction, in the context of machine learning, refers to the process of estimating or forecasting outcome or target variable based on available data. It models continuous valued functions by finding a mathematical relationship between input features and the target variable to make predictions. On the other hand, classification is the process of categorizing data points into predefined classes or labels in the discrete form based on their features. It is used to make decisions and assign objects to different categories or groups. In both prediction and classification tasks, the following steps are typically involved:

- Data collection: Gathering relevant data including input features and the target variable is the first step for prediction or classification tasks. It should involve systematically acquiring data from various sources, ensuring its accuracy and reliability, and organizing it for further use.
- Data preprocessing: It involves removing noise, transforming, organizing raw data into a format that is suitable for analysis, and splitting data into training and testing sets.
- Model selection: This involves selecting the suitable machine learning model considering factors like nature of the task (e.g., prediction or classification) and the specific characteristics of the data.

- Model training: In this step, training data is used to train the selected model. For classification, it includes learning the decision boundaries between different classes. For prediction, it involves finding the best-fitting model parameters. The training process may require several iterations for parameter tuning and validation to achieve the best possible model performance.
- Model evaluation: It helps to assess how well a trained machine learning model performs on new, unseen data by measuring the accuracy, mean squared error, etc. Cross-validation is often used to validate the model.
- Model deployment: It involves implementing the trained model in a real-world application to make predictions or classifications on new, unseen data.

The k-fold cross-validation and leave-one-out cross-validation are the widely used techniques in machine learning to assess the performance of a model and estimate its generalization error. In the k-fold cross-validation technique, initial data are divided into k mutually exclusive folds or subsets, each of approximately equal size. The model is trained and tested k times, each time using $(k-1)$ folds as the training set and the remaining fold as the testing set. The training and testing accuracies are averaged to evaluate the model's performance. In leave-one-out cross-validation technique, one data point is left out for testing while the rest are used for training. This process repeats for each data point, and the results are averaged.

In the realm of machine learning, data mining is an essential step, which involves exploring and analyzing data to extract hidden patterns from large datasets that can be used for prediction, classification, or knowledge discovery. Data mining and machine learning are complementing each other. Data mining focuses on exploratory data analysis and pattern discovery, while machine learning is concerned with developing algorithms that can make predictions or classifications based on data. Together, they form a powerful toolbox for extracting insights and creating predictive models from large and complex datasets.

1.2 Organization of the book

This book is divided into seven chapters, which discusses various AI techniques and their applications in the domain of textile engineering. A brief view of these chapters is given below.

This chapter is an elementary overview of AI. A brief idea of AI, machine learning, supervised learning, prediction, classification, cross-validation, and data mining is given in this chapter.

Artificial neural network (ANN) has emerged as a powerful and versatile tool in the field of AI and machine learning. An ANN emulates the nervous network system in humans and processes impulses in the same way as we do. Basic processing elements of an ANN are artificial neurons or nodes, which perform as summing and nonlinear mapping junctions. The artificial neurons are organized in layers and interconnected. The connection strength is known as weight, a numerical value constantly adjusted during learning to reach the desired goal. The total input is calculated by the summation of the weighted inputs. The sum is then passed through a transfer function, and subsequently the output is transmitted to each neuron of the next layer. A network with one or more hidden layers sandwiched between input and output layers can effectively approximate any nonlinear function and the network thus formed is called a multilayer ANN. An ANN learns from the historical training data. During the learning process, the predicted output is compared with the desired output, and an error signal is generated. The error signal is then minimized in iterative steps by adjusting the weights using a suitable training algorithm. The gradient descent backpropagation algorithm is the most popular learning method for ANN. Chapter 2 provides a comprehensive overview of ANN, delving into its foundational concepts, architecture, training methods, step by step working principle with MATLAB® coding, and practical applications in the field of textile engineering.

Chapter 3 deals with the SVM. An SVM stands as a prominent and versatile machine learning technique renowned for its efficacy in solving both binary and multiclass classification problems. Moreover, it is also convenient for prediction tasks. SVM is a machine learning system that uses a hypothesis space of linear functions in a high-dimensional feature space. It uses a kernel trick that transforms the original training data into a higher dimension feature space by a nonlinear mapping. It is trained with a learning algorithm based on a quadratic programming problem. The aim of support vector classification is to devise an optimal separating hyperplane in a high-dimensional feature space that separates data points linearly by maximizing the distance between different classes. Support vector regression performs prediction tasks by finding a hyperplane that best fits the data while minimizing the prediction error. This chapter explores basic principles of support vector classification and support vector regression in steps with illustrative examples, MATLAB coding, and applications in the textile domain.

Chapter 4 is designed to guide readers through the fundamentals of fuzzy logic, exploring its theoretical underpinnings and practical applications. In the realm of decision-making and problem-solving, traditional binary logic has long served as the cornerstone, relying on precise, clear-cut distinctions between true and false, on and off. However, many real-world situations are far from black and white, often residing in the vast shades of uncertainty, ambiguity, and imprecision. It is here that the paradigm of fuzzy logic emerges, providing a framework for handling the inherent fuzziness present in various human-centric and complex systems. Fuzzy logic acknowledges the nuances and imprecision inherent in human language and perception. Unlike classical logic, which deals with absolute truth values (0 or 1), fuzzy logic allows for degrees of truth between 0 and 1. This nuanced approach mirrors the human capacity to reason in shades of gray, making it a powerful tool for modeling systems where ambiguity is the norm rather than the exception. The foundation of fuzzy logic rests on several key concepts, including fuzzy sets, fuzzy membership functions, and fuzzy rules. Fuzzy sets generalize classical sets by assigning degrees of membership to elements, enabling a more flexible representation of uncertainty. Membership functions define the degree to which an element belongs to a fuzzy set, and fuzzy rules establish the relationships between fuzzy sets, providing a mechanism for mapping input to output in a fuzzy system. Fuzzy logic finds application in a wide array of fields, ranging from control systems and AI to decision support systems and pattern recognition. Its ability to model and manage uncertainty makes it particularly well-suited for tasks where traditional binary logic falls short. This chapter explains the mathematics of fuzzy sets, principles of fuzzy logic, and its potential to address complex, uncertain, and dynamic systems in the textile domain.

Chapter 5 delves into the realm of rough set theory, which also addresses uncertainty and vagueness in data such as fuzzy logic. Fuzzy logic deals with uncertainty by assigning degrees of membership to elements, whereas rough set theory focuses on approximating and classifying data with imprecise boundaries by discerning essential and nonessential attributes. The central idea behind rough set theory is to represent and analyze uncertainty in data by dividing it into two subsets such as lower approximation and upper approximation. These sets help define boundaries within which the information is certain or uncertain. The theory is especially valuable in situations where data is incomplete, imprecise, or inconsistent. This chapter begins with an exploration of the foundational concepts and mathematical tools used in rough set theory. This includes discussions on equivalence relations, approximation spaces, and discernibility matrices. As the chapter progresses, it explores the practical applications of rough set theory in the textile domain.

Chapter 6 covers nature-inspired optimization techniques, which are the computational methods that draw inspiration from natural processes and phenomena to solve complex optimization problems. The basic concepts of some popularly known nature-inspired optimization algorithms such as genetic algorithm that mimics the principles of evolution, particle swarm optimization that is inspired by social behavior of bird flocking, and ant colony optimization that imitates the ants' capability of finding the shortest route from the nest to a food source, are discussed in step by step with demonstrative examples and MATLAB coding. These techniques iteratively modify potential solutions, imitate the natural selection or social interactions among entities, and eventually converge toward the optimum solution. The potential applications of these optimization techniques in the domain of textile engineering are also discussed in this chapter.

Chapter 7 delves into the hybrid AI systems. The fundamental principles of two popularly used hybrid AI systems, namely adaptive neuro-fuzzy inference system and genetic fuzzy expert system have been discussed in step by step with suitable examples and MATLAB coding. The adaptive neuro-fuzzy inference system combines the adaptive learning capabilities of neural networks with the interpretability of fuzzy logic, whereas the genetic fuzzy expert system employs genetic algorithms to optimize the performance of the fuzzy expert system. The applications of these hybrid AI systems in the textile domain have also been discussed in this chapter.

Chapter 2

Artificial neural network

2.1 Introduction

Artificial neural network (ANN) is one of the popular soft computing techniques, which is inspired by the structure and function of the biological brain's neural networks. It is a fundamental component of modern artificial intelligence and machine learning. The foundation of ANN dates back to the work of McCulloch and Pitts (1943), who developed a simplified mathematical model of a neuron, which formed the basis for later developments in ANN. Later, Rosenblatt (1958) introduced the perceptron, a single-layer neural network capable of learning and making binary classifications, which revolutionized neural network research and laid the groundwork for future developments (Haykin, 2004; Kartalopoulos, 2000; Khanna, 1990; Yegnanarayana, 2009; Zurada, 2003). ANN is defined as a data processing system comprising highly interconnected simple neural computing elements that have the ability to learn and acquire knowledge to solve complex problems by simulating the behavior of its interconnected neurons. Each connection is expressed in terms of synaptic weights, which signifies the importance of the connection. Each neuron of an ANN takes multiple inputs, calculates its weighted sum by multiplying with synaptic weights, compares the weighted sum with a threshold value, applies a mathematical operation to it through an activation function, and produces an output. Basically, the activation function introduces nonlinearity into the system. The log-sigmoid function and hyperbolic tangent function are two widely used activation functions. A multilayer ANN consists of an input layer, one or more hidden layers, and an output layer. The most common ANN architecture is the feedforward backpropagation neural network, where information flows in the forward direction, from the input layer to hidden layers, and finally to the output layer, whereas error between the predicted output and the desired output is propagating backward through the network to update the weights and biases. During training, an ANN learns by adjusting the weights and biases of its neurons which keep on modified as the learning process continues.

ANN is extremely useful for cases in which a processing algorithm or analytical solutions are hard to find. It has been extensively utilized in fields necessitating the modeling of complex processes, extracting intricate patterns, and discerning relationships within large datasets. It can be used for a wide range of tasks, including pattern recognition, classification, and regression.

Artificial Intelligence in Textile Engineering. https://doi.org/10.1016/B978-0-443-15395-2.00008-9

2.2   ANN

ANN was developed in an attempt to imitate the functional principles of the human brain. The fundamental unit of the brain is the neuron, whose structure is illustrated in Fig. 2.1.

The axon transmits the neuron's output. Branches of the axon interface with the dendrites of other neurons through synapses. The signals from other neurons transmit down the dendrites to the cell body. The transmission of signals from one cell to another at a synapse is a complex electrochemical process that either raises or lowers the electrical potential of the body of the receiving cell. Once this potential reaches a threshold, a pulse is transmitted along the axon, and the cell is then said to have fired. It takes a millisecond (10^{-3} s) time scale to complete neural events, whereas events on an integrated circuit chip occur at the nanosecond (10^{-9} s) or picosecond (10^{-12} s) time scale. Although biological neurons are much slower than silicon chips, a human brain routinely accomplishes complex voice and pattern recognition tasks that would be the envy of a supercomputer. This is attributable to the fact that the human brain has about 100 billion neurons and around 60 trillion synapses or connections. This massive degree of parallelism makes up for the slowness of individual neurons. Thus, to design a system that can imitate a human brain, it is necessary to define a simple unit (i.e., an artificial neuron), join a number of such units through connections or weights, and allow these weights to decide the manner in which data is transferred from one unit to another. The whole system should be allowed to learn from examples by changing the weights iteratively. Once trained, the system will be capable of delivering an output for a given set of input parameters.

An artificial neuron is an attempt to capture the functional principles of a biological neuron. The schematic representation of the ANN model of McCulloch and Pitts (1943) is shown in Fig. 2.2.

In a simple neuron as shown in Fig. 2.2, $x_1, x_2, x_3, ..., x_n$ are the n numbers of inputs to the artificial neuron and $w_1, w_2, ..., w_n$ are the weights attached to the input links. Weights are actually multiplicative factors assigned to each input

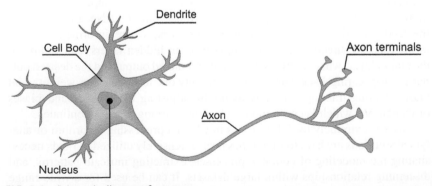

FIG. 2.1   Schematic diagram of a neuron.

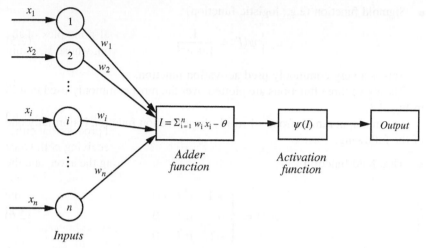

FIG. 2.2 Schematic diagram of a simple artificial neuron.

link to signify the contribution of each input to the total input received by the input layer. The summation of weighted inputs is compared with a threshold θ and the net weighted input becomes

$$I = \sum_{i=1}^{n} w_i x_i - \theta \tag{2.1}$$

Next, this calculated input (I) is passed through the activation function to produce the output given by

$$y = \psi(I) \tag{2.2}$$

where $\psi(\cdot)$ is a nonlinear function which is termed as activation or transfer function. For an output to lie between 0 and 1, popular choices of the activation function include the following:

- Threshold function

$$\psi(I) = \begin{cases} 1 & \text{if } I \geq 0 \\ 0 & \text{if } I < 0 \end{cases} \tag{2.3}$$

- Piecewise linear function

$$\psi(I) = \begin{cases} 1 & \text{if } I \geq 0.5 \\ I + 0.5 & \text{if } 0.5 > I > -0.5 \\ 0 & \text{if } I < -0.5 \end{cases} \tag{2.4}$$

- Sigmoid function (e.g., logistic function)

$$\psi(I) = \frac{1}{1 + e^{-I}} \qquad (2.5)$$

This is a very commonly used activation function.

The above three functions are plotted over the most commonly used ranges in Fig. 2.3.

For an output ranges from -1 to $+1$, the activation function is normally one of the following:

- Threshold function

$$\psi(I) = \begin{cases} +1 & \text{if } I > 0 \\ 0 & \text{if } I = 0 \\ -1 & \text{if } I < 0 \end{cases} \qquad (2.6)$$

- Sigmoid function (e.g., hyperbolic tangent function)

$$\psi(I) = \tanh(I) = \frac{e^I - e^{-I}}{e^I + e^{-I}} \qquad (2.7)$$

The two functions are plotted over the most commonly used ranges in Fig. 2.4.

The simplest ANN consists of two layers of neurons (Fig. 2.5). The introduction of more layers of neurons (known as hidden layers) between the input and output layers results in a multilayer network (Fig. 2.6).

A single-layer network can perform many simple operations. Logical operations are the building blocks of many complicated functions. The logical "AND" operator is represented in Table 2.1. One can see that when both the inputs are one, the output is one. In the other three cases, when either one or both inputs are zero, the output is zero. A graphical representation of the AND operation is depicted in Fig. 2.7 which shows that the dataset is linearly

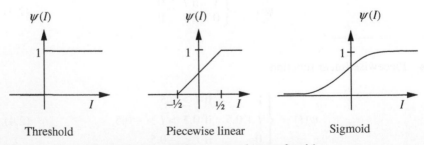

FIG. 2.3 Three activation functions to give an output between 0 and 1.

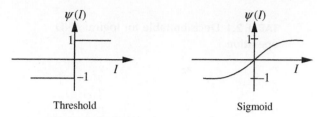

FIG. 2.4   Two activation functions to give an output between −1 and +1.

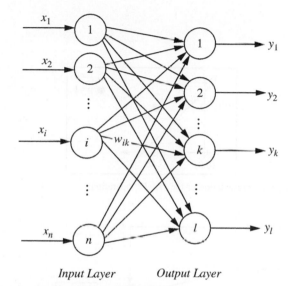

FIG. 2.5   A neural network without any hidden layer.

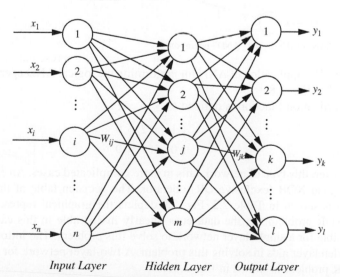

FIG. 2.6   A neural network with one hidden layer.

TABLE 2.1 Decision table for logical AND operation.

x_1	x_2	y
0	0	0
0	1	0
1	0	0
1	1	1

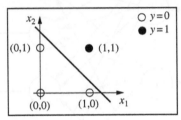

FIG. 2.7 A graphical representation of the "AND" operation.

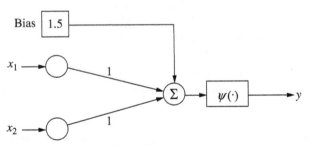

FIG. 2.8 ANN for the logical "AND" operation.

separable. The application of ANN for performing the logical AND operation is illustrated in Fig. 2.8.

The activation function is defined as

$$\psi(I) = \begin{cases} 1 & \text{if } I > 0 \\ 0 & \text{if } I \leq 0 \end{cases}$$

However, this type of network fails in more complicated cases. An example is the logical XOR (exclusive OR) operator. The decision table of the XOR problem is shown in Table 2.2. Fig. 2.9 depicts the graphical representation of the XOR problem. As the dataset is linearly inseparable in this case, it is not possible for a single-layer network to solve this problem. The introduction of a hidden layer aids in solving this problem. A two-layer network for solving the XOR problem is shown in Fig. 2.10.

TABLE 2.2 XOR problem.

x_1	x_2	y
1	1	0
0	1	1
1	0	1
0	0	0

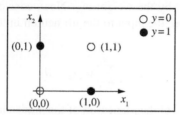

FIG. 2.9 A graphical representation of the "XOR" operation.

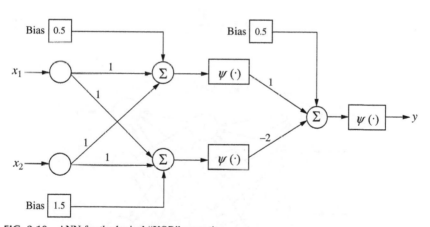

FIG. 2.10 ANN for the logical "XOR" operation.

From the above two examples of logical operations, an inference can be drawn that a simple neuron can able to model a linearly separable dataset but is incapable of modeling a nonlinearly separable dataset. A multilayer structure of neurons is required for a linearly inseparable dataset. Kolmogorov's theorem states that any continuous function $f(x_1, x_2, ..., x_n)$ of n variables $x_1, x_2, ..., x_n$ can be represented in the form

$$f(x_1, x_2, ..., x_n) = \sum_{j=1}^{2n+1} h_j \left(\sum_{i=1}^{n} g_{ij}(x_i) \right) \qquad (2.8)$$

where h_j and g_{ij} are continuous functions of one variable and g_{ij}'s are fixed monotonically increasing functions. In general, according to Kolmogorov's theorem, multilayer networks should be capable of approximating any function to any degree of accuracy.

A three-layer feedforward backpropagation ANN is depicted in Fig. 2.11 where the indices i, j, and k refer to the neurons in the input, hidden, and output layers, respectively. The input signals x_1, x_2, ..., x_n, are propagated through the network in the forward direction (from left to right), whereas the error signals e_1, e_2, ..., e_l are propagated in the backward direction (from right to left). The ith neuron in the input layer and jth neuron in the hidden layer are connected by the weight w_{ij} and jth neuron in the hidden layer and kth neuron in the output layer are connected by the weight w_{jk}. Now, considering the logistic sigmoid activation function, the input to the jth neuron in the hidden layer can be calculated as

$$I_j = \sum_{i=1}^{n} w_{ij}x_i - \theta_j \qquad (2.9)$$

where θ_j is the threshold of the jth neuron of the hidden layer. The output of the jth hidden neuron is given by

$$y_j = \frac{1}{1 + e^{-I_j}} \qquad (2.10)$$

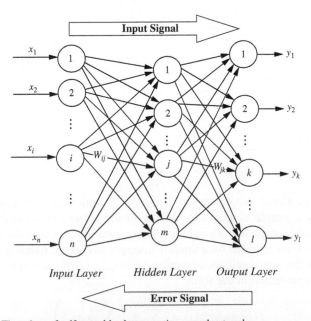

FIG. 2.11   Three-layer feedforward backpropagation neural network.

The input to the output neurons is the weighted sum of the outputs of the hidden neurons. Therefore,

$$I_k = \sum_{j=1}^{m} w_{jk} y_j - \theta_k \qquad (2.11)$$

where θ_k is the threshold of the kth neuron of the output layer. The output of the kth output neuron is given by

$$y_k = \frac{1}{1 + e^{-I_k}} \qquad (2.12)$$

Neural networks learn from examples. Typically, many input-output pairs are used to train a network. In supervised learning a comparison is made between the network's computed output and the correct expected output to determine the learning error. The error, so determined, is then estimated to modify network parameters to improve the performance of the network. The sum squared error over all the output units for pth training pattern is given by

$$E_p = \frac{1}{2} \sum_{k=1}^{l} (y_{d,k} - y_k)^2 \qquad (2.13)$$

where $y_{d,k}$ and y_k are the desired output and calculated output, respectively, for the kth neuron. The sum square error over the N training patterns is calculated as

$$E = \sum_{p=1}^{N} E_p$$
$$= \frac{1}{2} \sum_{p=1}^{N} \sum_{k=1}^{l} (y_{d,k} - y_k)^2 \qquad (2.14)$$

The sum square error (E) is a useful indicator for the performance of a network. The backpropagation training algorithm attempts to minimize this error signal. In this algorithm, the error signal is propagated backward to the neural network, and the weights are adjusted using the gradient descent method. In this manner, the error signal decreases in epoch after epoch. When the value of the total sum square error in an entire pass through all training sets, or epoch, becomes sufficiently small, a network can be considered to be a trained one. The weight-updating process of the feedforward backpropagation network is explained below.

The propagation of error signals starts from the output layer. The error signal of the kth neuron in the output layer at iteration t is given by

$$e_k(t) = y_{d,k}(t) - y_k(t) \qquad (2.15)$$

The weights of the output layer are updated as

$$w_{jk}(t+1) = w_{jk}(t) + \Delta w_{jk}(t) \qquad (2.16)$$

where $\Delta w_{jk}(t)$ is the weight correction in the output layer. Let $y_j(t)$ be the output of the jth neuron in the hidden layer at iteration t. The weight correction in the output layer, $\Delta w_{jk}(t)$ is estimated by

$$\Delta w_{jk}(t) = \alpha y_j(t)\delta_k(t) \tag{2.17}$$

where $\delta_k(t)$ is the error gradient of kth neuron in the output layer at iteration t and α is the learning rate which lies between $0 < \alpha < 1$. The error gradient, $\delta_k(t)$ is expressed as

$$\delta_k(t) = \frac{\partial y_k(t)}{\partial I_k(t)} e_k(t) \tag{2.18}$$

where $y_k(t)$ is the output of kth neuron in the output layer at iteration t and $I_k(t)$ is the net weighted input to the same neuron at the same iteration. In case of a logistic sigmoid activation function, we have

$$y_k(t) = \frac{1}{1 + e^{-I_k(t)}} \tag{2.19}$$

Thus, Eq. (2.18) becomes

$$
\begin{aligned}
\delta_k(t) &= \frac{\partial\left\{\dfrac{1}{1 + e^{-I_k(t)}}\right\}}{\partial I_k(t)} e_k(t) \\
&= \frac{e^{-I_k(t)}}{\{1 + e^{-I_k(t)}\}^2} e_k(t) \\
&= \left\{\frac{1}{1 + e^{-I_k(t)}}\right\}\left\{1 - \frac{1}{1 + e^{-I_k(t)}}\right\} e_k(t) \\
&= y_k(t)\{1 - y_k(t)\} e_k(t)
\end{aligned}
\tag{2.20}
$$

Similarly, the weight correction in the hidden layer, $\Delta w_{ij}(t)$ can be estimated as

$$\Delta w_{ij}(t) = \alpha x_i \delta_j(t) \tag{2.21}$$

where $\delta_j(t)$ is the error gradient of jth neuron in the hidden layer at iteration t. The error gradient, $\delta_j(t)$ can be expressed as

$$
\begin{aligned}
\delta_j(t) &= \frac{\partial\left\{\dfrac{1}{1 + e^{-I_j(t)}}\right\}}{\partial I_j(t)} \sum_{k=1}^{l} \delta_k(t)\Delta w_{jk}(t) \\
&= \frac{e^{-I_j(t)}}{\{1 + e^{-I_j(t)}\}^2} \sum_{k=1}^{l} \delta_k(t)\Delta w_{jk}(t) \\
&= \left\{\frac{1}{1 + e^{-I_j(t)}}\right\}\left\{1 - \frac{1}{1 + e^{-I_j(t)}}\right\} \sum_{k=1}^{l} \delta_k(t)\Delta w_{jk}(t) \\
&= y_j(t)\{1 - y_j(t)\} \sum_{k=1}^{l} \delta_k(t)\Delta w_{jk}(t)
\end{aligned}
$$

$$\tag{2.22}$$

where

$$I_j(t) = \sum_{i=1}^{n} w_{ij}(t)x_i - \theta_j(t) \qquad (2.23)$$

The weights of the hidden layer are updated as

$$w_{ij}(t+1) = w_{ij}(t) + \Delta w_{ij}(t) \qquad (2.24)$$

A low learning rate (α) may cause the network to converge slowly, which can be time-consuming for large datasets, but it may also result in a more accurate model. On the other hand, a high learning rate can cause the network to converge quickly, but it may also cause the weights to oscillate around the optimal values and thus preventing the network from converging to the optimal solution. Nevertheless, the training of the network can be accelerated by including a momentum term in Eq. (2.17) as follows:

$$\Delta w_{jk}(t) = \beta \Delta w_{jk}(t-1) + \alpha y_j(t)\delta_k(t) \qquad (2.25)$$

where β is the momentum term which lies between $0 \leq \beta < 1$.

2.3 Step-by-step working principle of ANN

To demonstrate the step-by-step working principle of ANN, let us consider the Exclusive OR (XOR) problem as given in Table 2.2. A multilayer feedforward neural network is depicted in Fig. 2.12.

The initial weights and threshold levels are randomly selected as follows:

$$w_{13} = 0.1, w_{23} = 0.2, w_{14} = 0.5, w_{24} = 0.8, w_{35} = 1, w_{45} = -1, \theta_3 = 0.4,$$

$$\theta_4 = 0.2 \text{ and } \theta_5 = 0.5.$$

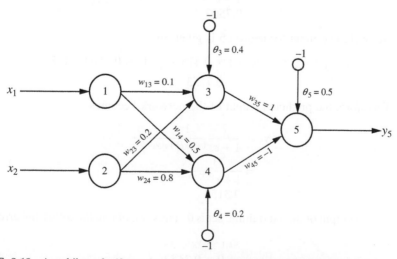

FIG. 2.12 A multilayer feedforward neural network with initial weight and bias values.

In the given example, we have four training sets, thus each epoch is comprised of four iterations. A step-by-step calculation of updating all the weights and the threshold levels in the 1st epoch with four iterations is shown below.

1st iteration, epoch 1:

For the 1st training set, we have

$$x_1 = 1, x_2 = 1 \text{ and } y_d = 0.$$

Net weighted input for neuron 3 is thus given by

$$w_{13}x_1 + w_{23}x_2 - \theta_3 = 0.1 \times 1 + 0.2 \times 1 - 0.4$$
$$= -0.1$$

Hence, the output of neuron 3 is estimated as

$$y_3 = \frac{1}{1 + e^{-(w_{13}x_1 + w_{23}x_2 - \theta_3)}}$$
$$= \frac{1}{1 + e^{0.1}}$$
$$= 0.475$$

Net weighted input for neuron 4 is given by

$$w_{14}x_1 + w_{24}x_2 - \theta_4 = 0.5 \times 1 + 0.8 \times 1 - 0.2$$
$$= 1.1$$

Thus, the output of neuron 4 becomes

$$y_4 = \frac{1}{1 + e^{-(w_{14}x_1 + w_{24}x_2 - \theta_4)}}$$
$$= \frac{1}{1 + e^{-1.1}}$$
$$= 0.7503$$

Net weighted input for neuron 5 is given by

$$w_{35}y_3 + w_{45}y_4 - \theta_5 = 1 \times 0.475 + (-1) \times 0.7503 - 0.5$$
$$= -0.7753$$

Therefore, the predicted output of the network becomes

$$y_5 = \frac{1}{1 + e^{-(w_{35}y_3 + w_{45}y_4 - \theta_5)}}$$
$$= \frac{1}{1 + e^{0.7753}}$$
$$= 0.3153$$

Actual output of the 1st training set is 0. Thus, we obtain the following error:

$$e_{k,1} = y_d - y_5$$
$$= 0 - 0.3153$$
$$= -0.3153$$

The error gradient for neuron 5 can be calculated as follows:

$$\delta_5 = y_5(1 - y_5)e_{k,1}$$
$$= 0.3153 \times (1 - 0.3153) \times (-0.3153)$$
$$= -0.0681$$

By assuming the learning rate parameter $\alpha = 0.1$, the corrections of weights and threshold levels which are connected with neuron 5 become

$$\Delta w_{35} = \alpha y_3 \delta_5$$
$$= 0.1 \times 0.475 \times (-0.0681)$$
$$= -0.0032$$

$$\Delta w_{45} = \alpha y_4 \delta_5$$
$$= 0.1 \times 0.7503 \times (-0.0681)$$
$$= -0.0051$$

$$\Delta \theta_5 = \alpha(-1)\delta_5$$
$$= 0.1 \times (-1) \times (-0.0681)$$
$$= 0.0068$$

The error gradients for neuron 3 and neuron 4 become

$$\delta_3 = y_3(1 - y_3)\delta_5 w_{35}$$
$$= 0.475 \times (1 - 0.475) \times (-0.0681) \times 1$$
$$= -0.017$$

$$\delta_4 = y_4(1 - y_4)\delta_5 w_{45}$$
$$= 0.7503 \times (1 - 0.7503) \times (-0.0681) \times (-1)$$
$$= 0.0128$$

The corrections of weights and threshold levels which are connected with neuron 3 and neuron 4 of hidden layers are estimated as

$$\Delta w_{13} = \alpha x_1 \delta_3$$
$$= 0.1 \times 1 \times (-0.017)$$
$$= -0.0017$$

$$\Delta w_{23} = \alpha x_2 \delta_3$$
$$= 0.1 \times 1 \times (-0.017)$$
$$= -0.0017$$

$$\Delta w_{14} = \alpha x_1 \delta_4$$
$$= 0.1 \times 1 \times 0.0128$$
$$= 0.0013$$

$$\Delta w_{24} = \alpha x_2 \delta_4$$
$$= 0.1 \times 1 \times 0.0128$$
$$= 0.0013$$

$$\Delta \theta_3 = \alpha(-1)\delta_3$$
$$= 0.1 \times (-1) \times (-0.017)$$
$$= 0.0017$$

$$\Delta \theta_4 = \alpha(-1)\delta_4$$
$$= 0.1 \times (-1) \times 0.0128$$
$$= -0.0013$$

Thus, the weights and threshold levels are updated as follows:

$$w_{13} = w_{13} + \Delta w_{13}$$
$$= 0.1 + (-0.0017)$$
$$= 0.0983$$

$$w_{23} = w_{23} + \Delta w_{23}$$
$$= 0.2 + (-0.0017)$$
$$= 0.1983$$

$$w_{14} = w_{14} + \Delta w_{14}$$
$$= 0.5 + 0.0013$$
$$= 0.5013$$

$$w_{24} = w_{24} + \Delta w_{24}$$
$$= 0.8 + 0.0013$$
$$= 0.8013$$

$$w_{35} = w_{35} + \Delta w_{35}$$
$$= 1 + (-0.0032)$$
$$= 0.9968$$

$$w_{45} = w_{45} + \Delta w_{45}$$
$$= -1 + (-0.0051)$$
$$= -1.0051$$

$$\theta_3 = \theta_3 + \Delta \theta_3$$
$$= 0.4 + 0.0017$$
$$= 0.4017$$

$$\theta_4 = \theta_4 + \Delta\theta_4$$
$$= 0.2 + (-0.0013)$$
$$= 0.1987$$
$$\theta_5 = \theta_5 + \Delta\theta_5$$
$$= 0.5 + 0.0068$$
$$= 0.5068$$

Fig. 2.13 shows the updated weights and biases after 1st iteration.

2nd iteration, epoch 1:
For the 2nd training set, we have

$$x_1 = 0, x_2 = 1 \text{ and } y_d = 1.$$

Net weighted input for neuron 3 is thus given by

$$w_{13}x_1 + w_{23}x_2 - \theta_3 = 0.0983 \times 0 + 0.1983 \times 1 - 0.4017$$
$$= -0.2034$$

Hence, the output of neuron 3 is estimated as

$$y_3 = \frac{1}{1 + e^{-(w_{13}x_1 + w_{23}x_2 - \theta_3)}}$$
$$= \frac{1}{1 + e^{0.2034}}$$
$$= 0.4493$$

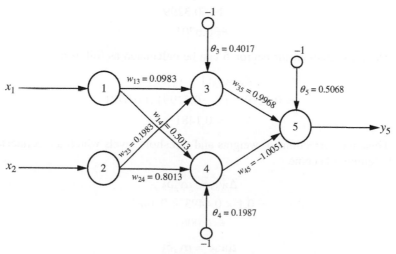

FIG. 2.13 Neural network with adjusted weight and bias values after 1st iteration.

Net weighted input for neuron 4 is given by

$$w_{14}x_1 + w_{24}x_2 - \theta_4 = 0.5013 \times 0 + 0.8013 \times 1 - 0.1987$$
$$= 0.6026$$

Thus, the output of neuron 4 becomes

$$y_4 = \frac{1}{1 + e^{-(w_{14}x_1 + w_{24}x_2 - \theta_4)}}$$
$$= \frac{1}{1 + e^{-0.6026}}$$
$$= 0.6462$$

Net weighted input for neuron 5 is given by

$$w_{35}y_3 + w_{45}y_4 - \theta_5 = 0.9968 \times 0.4493 + (-1.0051) \times 0.6462 - 0.5068$$
$$= -0.7085$$

Therefore, the predicted output of the network becomes

$$y_5 = \frac{1}{1 + e^{-(w_{35}y_3 + w_{45}y_4 - \theta_5)}}$$
$$= \frac{1}{1 + e^{0.7085}}$$
$$= 0.3299$$

Actual output of the 2nd training set is 1. Thus, we obtain the following error:

$$e_{k,2} = y_d - y_5$$
$$= 1 - 0.3299$$
$$= 0.6701$$

The error gradient for neuron 5 can be calculated as follows:

$$\delta_5 = y_5(1 - y_5)e_{k,2}$$
$$= 0.3299 \times (1 - 0.3299) \times 0.6701$$
$$= 0.1481$$

Thus, the corrections of weights and threshold levels which are connected with neuron 5 become

$$\Delta w_{35} = \alpha y_3 \delta_5$$
$$= 0.1 \times 0.4493 \times 0.1481$$
$$= 0.0067$$

$$\Delta w_{45} = \alpha y_4 \delta_5$$
$$= 0.1 \times 0.6462 \times 0.1481$$
$$= 0.0096$$

$$\Delta\theta_5 = \alpha(-1)\delta_5$$
$$= 0.1 \times (-1) \times 0.1481$$
$$= -0.0148$$

The error gradients for neuron 3 and neuron 4 become

$$\delta_3 = y_3(1 - y_3)\delta_5 w_{35}$$
$$= 0.4493 \times (1 - 0.4493) \times 0.1481 \times 0.9968$$
$$= 0.0365$$

$$\delta_4 = y_4(1 - y_4)\delta_5 w_{45}$$
$$= 0.6462 \times (1 - 0.6462) \times 0.1481 \times (-1.0051)$$
$$= -0.034$$

The corrections of weights and threshold levels which are connected with neuron 3 and neuron 4 of hidden layers are estimated as

$$\Delta w_{13} = \alpha x_1 \delta_3$$
$$= 0.1 \times 0 \times 0.0365$$
$$= 0$$

$$\Delta w_{23} = \alpha x_2 \delta_3$$
$$= 0.1 \times 1 \times 0.0365$$
$$= 0.0037$$

$$\Delta w_{14} = \alpha x_1 \delta_4$$
$$= 0.1 \times 0 \times (-0.034)$$
$$= 0$$

$$\Delta w_{24} = \alpha x_2 \delta_4$$
$$= 0.1 \times 1 \times (-0.034)$$
$$= -0.0034$$

$$\Delta\theta_3 = \alpha(-1)\delta_3$$
$$= 0.1 \times (-1) \times 0.0365$$
$$= -0.0037$$

$$\Delta\theta_4 = \alpha(-1)\delta_4$$
$$= 0.1 \times (-1) \times (-0.034)$$
$$= 0.0034$$

Thus, the weights and threshold levels are updated as follows:

$$w_{13} = w_{13} + \Delta w_{13}$$
$$= 0.0983 + 0$$
$$= 0.0983$$

$$w_{23} = w_{23} + \Delta w_{23}$$
$$= 0.1983 + 0.0037$$
$$= 0.202$$

$$w_{14} = w_{14} + \Delta w_{14}$$
$$= 0.5013 + 0$$
$$= 0.5013$$

$$w_{24} = w_{24} + \Delta w_{24}$$
$$= 0.8013 + (-0.0034)$$
$$= 0.7979$$

$$w_{35} = w_{35} + \Delta w_{35}$$
$$= 0.9968 + 0.0067$$
$$= 1.0034$$

$$w_{45} = w_{45} + \Delta w_{45}$$
$$= -1.0051 + 0.0096$$
$$= -0.9955$$

$$\theta_3 = \theta_3 + \Delta \theta_3$$
$$= 0.4017 + (-0.0037)$$
$$= 0.398$$

$$\theta_4 = \theta_4 + \Delta \theta_4$$
$$= 0.1987 + 0.0034$$
$$= 0.2021$$

$$\theta_5 = \theta_5 + \Delta \theta_5$$
$$= 0.5068 + (-0.0148)$$
$$= 0.492$$

Fig. 2.14 depicts the updated weights and biases after 2nd iteration.

3rd iteration, epoch 1:
For the 3rd training set, we have

$$x_1 = 1, x_2 = 0, \text{and } y_d = 1.$$

Net weighted input for neuron 3 is thus given by

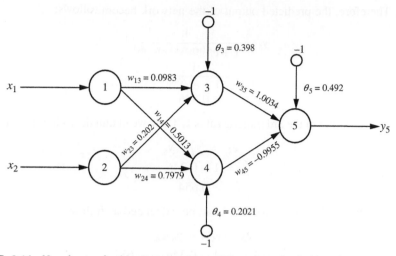

FIG. 2.14   Neural network with adjusted weight and bias values after 2nd iteration.

$$w_{13}x_1 + w_{23}x_2 - \theta_3 = 0.0983 \times 1 + 0.202 \times 0 - 0.398$$
$$= -0.2997$$

Hence, the output of neuron 3 is estimated as

$$y_3 = \frac{1}{1 + e^{-(w_{13}x_1 + w_{23}x_2 - \theta_3)}}$$
$$= \frac{1}{1 + e^{0.2997}}$$
$$= 0.4256$$

Net weighted input for neuron 4 is given by

$$w_{14}x_1 + w_{24}x_2 - \theta_4 = 0.5013 \times 1 + 0.7979 \times 0 - 0.2021$$
$$= 0.2991$$

Thus, the output of neuron 4 becomes

$$y_4 = \frac{1}{1 + e^{-(w_{14}x_1 + w_{24}x_2 - \theta_4)}}$$
$$= \frac{1}{1 + e^{-0.2991}}$$
$$= 0.5742$$

Net weighted input for neuron 5 is given by

$$w_{35}y_3 + w_{45}y_4 - \theta_5 = 1.0034 \times 0.4256 + (-0.9955) \times 0.5742 - 0.492$$
$$= -0.6366$$

Therefore, the predicted output of the network becomes

$$y_5 = \frac{1}{1 + e^{-(w_{35}y_3 + w_{45}y_4 - \theta_5)}}$$

$$= \frac{1}{1 + e^{0.6366}}$$

$$= 0.346$$

Actual output of the 3rd training set is 1. Thus, we obtain the following error:

$$e_{k,3} = y_d - y_5$$

$$= 1 - 0.346$$

$$= 0.654$$

The error gradient for neuron 5 can be calculated as follows:

$$\delta_5 = y_5(1 - y_5)e_{k,3}$$

$$= 0.346 \times (1 - 0.346) \times 0.654$$

$$= 0.148$$

Thus, the corrections of weights and threshold levels which are connected with neuron 5 become

$$\Delta w_{35} = \alpha y_3 \delta_5$$

$$= 0.1 \times 0.4256 \times 0.148$$

$$= 0.0063$$

$$\Delta w_{45} = \alpha y_4 \delta_5$$

$$= 0.1 \times 0.5742 \times 0.148$$

$$= 0.0085$$

$$\Delta \theta_5 = \alpha(-1)\delta_5$$

$$= 0.1 \times (-1) \times 0.148$$

$$= -0.0148$$

The error gradients for neuron 3 and neuron 4 become

$$\delta_3 = y_3(1 - y_3)\delta_5 w_{35}$$

$$= 0.4256 \times (1 - 0.4256) \times 0.148 \times 1.0034$$

$$= 0.0363$$

$$\delta_4 = y_4(1 - y_4)\delta_5 w_{45}$$

$$= 0.5742 \times (1 - 0.5742) \times 0.148 \times (-0.9955)$$

$$= -0.036$$

The corrections of weights and threshold levels which are connected with neuron 3 and neuron 4 of hidden layers are estimated as

$$\Delta w_{13} = \alpha x_1 \delta_3$$
$$= 0.1 \times 1 \times 0.0365$$
$$= 0.0036$$

$$\Delta w_{23} = \alpha x_2 \delta_3$$
$$= 0.1 \times 0 \times 0.0365$$
$$= 0$$

$$\Delta w_{14} = \alpha x_1 \delta_4$$
$$= 0.1 \times 1 \times (-0.036)$$
$$= -0.0036$$

$$\Delta w_{24} = \alpha x_2 \delta_4$$
$$= 0.1 \times 0 \times (-0.036)$$
$$= 0$$

$$\Delta \theta_3 = \alpha(-1)\delta_3$$
$$= 0.1 \times (-1) \times 0.0363$$
$$= -0.0036$$

$$\Delta \theta_4 = \alpha(-1)\delta_4$$
$$= 0.1 \times (-1) \times (-0.036)$$
$$= 0.0036$$

Thus, in the 3rd iteration, the weights and threshold levels are updated as follows:

$$w_{13} = w_{13} + \Delta w_{13}$$
$$= 0.0983 + 0.0036$$
$$= 0.1019$$

$$w_{23} = w_{23} + \Delta w_{23}$$
$$= 0.202 + 0$$
$$= 0.202$$

$$w_{14} = w_{14} + \Delta w_{14}$$
$$= 0.5013 + (-0.0036)$$
$$= 0.4977$$

$$w_{24} = w_{24} + \Delta w_{24}$$
$$= 0.7979 + 0$$
$$= 0.7979$$

$$w_{35} = w_{35} + \Delta w_{35}$$
$$= 1.0034 + 0.0063$$
$$= 1.0097$$

$$w_{45} = w_{45} + \Delta w_{45}$$
$$= -0.9955 + 0.0085$$
$$= -0.987$$

$$\theta_3 = \theta_3 + \Delta \theta_3$$
$$= 0.398 + (-0.0036)$$
$$= 0.3944$$

$$\theta_4 = \theta_4 + \Delta \theta_4$$
$$= 0.2021 + 0.0036$$
$$= 0.2057$$

$$\theta_5 = \theta_5 + \Delta \theta_5$$
$$= 0.492 + (-0.0148)$$
$$= 0.4772$$

Fig. 2.15 shows the adjusted weight and bias values after 3rd iteration.

4th iteration, epoch 1:
For the 4th training set, we have

$$x_1 = 0, x_2 = 0 \text{ and } y_d = 0.$$

Net weighted input for neuron 3 is thus given by

$$w_{13}x_1 + w_{23}x_2 - \theta_3 = 0.1019 \times 0 + 0.202 \times 0 - 0.3944$$
$$= -0.3944$$

Hence, the output of neuron 3 is estimated as

$$y_3 = \frac{1}{1 + e^{-(w_{13}x_1 + w_{23}x_2 - \theta_3)}}$$
$$= \frac{1}{1 + e^{0.3944}}$$
$$= 0.4027$$

Net weighted input for neuron 4 is given by

$$w_{14}x_1 + w_{24}x_2 - \theta_4 = 0.4977 \times 0 + 0.7979 \times 0 - 0.2057$$
$$= -0.2057$$

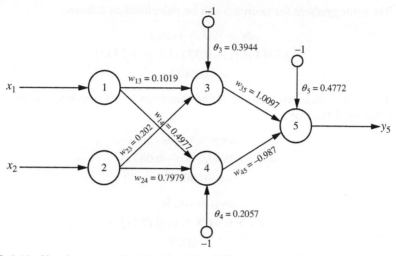

FIG. 2.15 Neural network with adjusted weight and bias values after 3rd iteration.

Thus, the output of neuron 4 becomes

$$y_4 = \frac{1}{1 + e^{-(w_{14}x_1 + w_{24}x_2 - \theta_4)}}$$
$$= \frac{1}{1 + e^{0.2057}}$$
$$= 0.4487$$

Net weighted input for neuron 5 is given by

$$w_{35}y_3 + w_{45}y_4 - \theta_5 = 1.0097 \times 0.4027 + (-0.987) \times 0.4487 - 0.4772$$
$$= -0.5136$$

Therefore, the predicted output of the network becomes

$$y_5 = \frac{1}{1 + e^{-(w_{35}y_3 + w_{45}y_4 - \theta_5)}}$$
$$= \frac{1}{1 + e^{0.5136}}$$
$$= 0.3744$$

Actual output of the 4th training set is 0. Thus, we obtain the following error:

$$e_{k,4} = y_d - y_5$$
$$= 0 - 0.3744$$
$$= -0.3744$$

The error gradient for neuron 5 can be calculated as follows:

$$\delta_5 = y_5(1 - y_5)e_{k,4}$$
$$= 0.3744 \times (1 - 0.3744) \times (-0.3744)$$
$$= -0.0877$$

Thus, the corrections of weights and threshold levels which are connected with neuron 5 become

$$\Delta w_{35} = \alpha y_3 \delta_5$$
$$= 0.1 \times 0.4027 \times (-0.0877)$$
$$= -0.0035$$

$$\Delta w_{45} = \alpha y_4 \delta_5$$
$$= 0.1 \times 0.4478 \times (-0.0877)$$
$$= -0.0039$$

$$\Delta \theta_5 = \alpha(-1)\delta_5$$
$$= 0.1 \times (-1) \times (-0.0877)$$
$$= 0.0088$$

The error gradients for neuron 3 and neuron 4 become

$$\delta_3 = y_3(1 - y_3)\delta_5 w_{35}$$
$$= 0.4027 \times (1 - 0.4027) \times (-0.0877) \times 1.0097$$
$$= -0.0213$$

$$\delta_4 = y_4(1 - y_4)\delta_5 w_{45}$$
$$= 0.4487 \times (1 - 0.4487) \times (-0.0877) \times (-0.987)$$
$$= 0.0214$$

The corrections of weights and threshold levels which are connected with neuron 3 and neuron 4 of hidden layers are estimated as

$$\Delta w_{13} = \alpha x_1 \delta_3$$
$$= 0.1 \times 0 \times (-0.0213)$$
$$= 0$$

$$\Delta w_{23} = \alpha x_2 \delta_3$$
$$= 0.1 \times 0 \times (-0.0213)$$
$$= 0$$

$$\Delta w_{14} = \alpha x_1 \delta_4$$
$$= 0.1 \times 0 \times 0.0214$$
$$= 0$$

$$\Delta w_{24} = \alpha x_2 \delta_4$$
$$= 0.1 \times 0 \times 0.0214$$
$$= 0$$

$$\Delta \theta_3 = \alpha(-1)\delta_3$$
$$= 0.1 \times (-1) \times (-0.0213)$$
$$= 0.0021$$

$$\Delta \theta_4 = \alpha(-1)\delta_4$$
$$= 0.1 \times (-1) \times 0.0214$$
$$= -0.0021$$

Thus, in the 4th iteration, the weights and threshold levels are updated as follows:

$$w_{13} = w_{13} + \Delta w_{13}$$
$$= 0.1019 + 0$$
$$= 0.1019$$

$$w_{23} = w_{23} + \Delta w_{23}$$
$$= 0.202 + 0$$
$$= 0.202$$

$$w_{14} = w_{14} + \Delta w_{14}$$
$$= 0.4977 + 0$$
$$= 0.4977$$

$$w_{24} = w_{24} + \Delta w_{24}$$
$$= 0.7979 + 0$$
$$= 0.7979$$

$$w_{35} = w_{35} + \Delta w_{35}$$
$$= 1.0097 + (-0.0035)$$
$$= 1.0062$$

$$w_{45} = w_{45} + \Delta w_{45}$$
$$= -0.987 + (-0.0039)$$
$$= -0.991$$

$$\theta_3 = \theta_3 + \Delta \theta_3$$
$$= 0.3944 + 0.0021$$
$$= 0.3965$$

$$\theta_4 = \theta_4 + \Delta\theta_4$$
$$= 0.2057 + (-0.0021)$$
$$= 0.2036$$

$$\theta_5 = \theta_5 + \Delta\theta_5$$
$$= 0.4772 + 0.0088$$
$$= 0.486$$

Fig. 2.16 shows the adjusted weight and bias values after 4th iteration. Table 2.3 shows the summary of calculations of network training for 1st epoch.

After completion of 1st epoch which comprises four iterations, the sum square error over the four training patterns becomes

$$E = \frac{1}{2} \sum_{p=1}^{4} e_{k,p}^2$$

$$= \frac{1}{2}(0.1032 + 0.4376 + 0.424 + 0.1379)$$

$$= 0.5514$$

In the backpropagation learning algorithm, the sum square error (E) is decreased in epoch after epoch. Fig. 2.17 depicts the learning curve up to 100 epochs or 400 iterations. In our example, the stopping criterion is defined as $E < 0.0005$. It took 59,829 epochs or 239,316 iterations to train the network

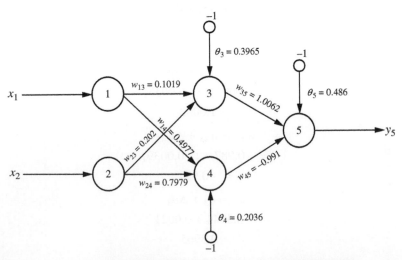

FIG. 2.16 Neural network with adjusted weight and bias values after 4th iteration.

TABLE 2.3 A summary of calculations of neural network training for 1st epoch.

	1st training set	2nd training set	3rd training set	4th training set
	$x_1 = 1,$ $x_2 = 1,$ $y_d = 0$	$x_1 = 0,$ $x_2 = 1,$ $y_d = 1$	$x_1 = 1,$ $x_2 = 0,$ $y_d = 1$	$x_1 = 0,$ $x_2 = 0,$ $y_d = 0$
$w_{13}x_1 + w_{23}x_2 - \theta_3$	−0.0927	−0.1946	−0.2946	−0.3965
$y_3 = \dfrac{1}{1 + e^{-(w_{13}x_1 + w_{23}x_2 - \theta_3)}}$	0.4769	0.4515	0.4269	0.4021
$w_{14}x_1 + w_{24}x_2 - \theta_4$	1.092	0.5943	0.2941	−0.2036
$y_4 = \dfrac{1}{1 + e^{-(w_{14}x_1 + w_{24}x_2 - \theta_4)}}$	0.7487	0.6443	0.573	0.4493
$w_{35}y_3 + w_{45}y_4 - \theta_5$	−0.7481	−0.6702	−0.6243	−0.5266
$y_5 = \dfrac{1}{1 + e^{-(w_{35}y_3 + w_{45}y_4 - \theta_5)}}$	0.3212	0.3385	0.3488	0.3713
y_d	0	1	1	0
$e_{k,p} = y_d - y_5$	−0.3212	0.6615	0.6512	−0.3713
$e_{k,p}^2 = (y_d - y_5)^2$	0.1032	0.4376	0.424	0.1379

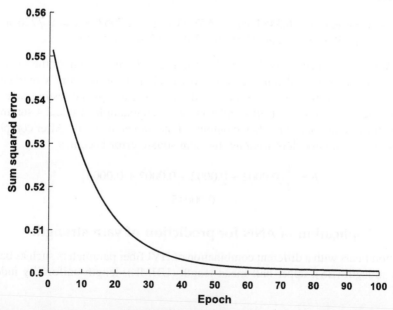

FIG. 2.17 Learning curve up to 100 epochs or 400 iterations.

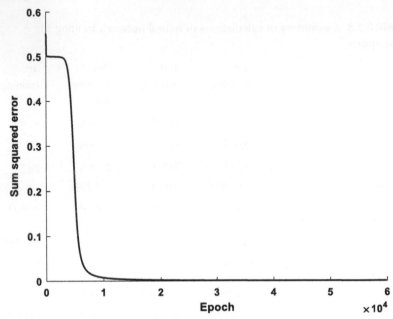

FIG. 2.18 The learning curve for the entire epochs for solving the XOR operation.

for solving XOR operation. The learning curve for all the epochs is illustrated in Fig. 2.18.

The final weights and threshold levels satisfying the chosen stopping criterion are calculated using a MATLAB® coding which gives the following results:

$$w_{13} = 6.3478, w_{23} = 6.3484, w_{14} = 4.7953, w_{24} = 4.7953, w_{35} = 9.7726, w_{45}$$
$$= -10.3536, \theta_3 = 2.8221, \theta_4 = 7.3566 \text{ and } \theta_5 = 4.5706.$$

With this trained network as depicted in Fig. 2.19, the testing of the same is done by presenting all training sets and calculating the network's predicted output. Here, it may be noted that we do not have a separate testing dataset as the training dataset is itself smaller in the given example. Table 2.4 summarizes the calculations of predicted outputs of the trained network. After completion of the final epoch of training, the sum square error becomes

$$E = \frac{1}{2}(0.0002 + 0.0002 + 0.0002 + 0.0003)$$
$$= 0.00045$$

2.4 Application of ANN for prediction of yarn strength

Cotton fibers with a different combination of HVI fiber parameters such as bundle strength (BS), upper half mean length (UHML), length uniformity index

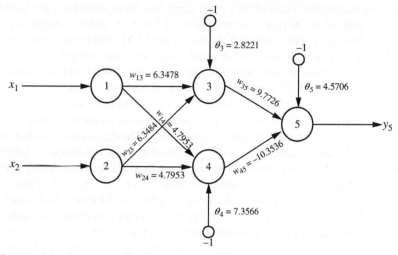

FIG. 2.19 Neural network with adjusted weight and bias values after completion of training.

TABLE 2.4 A summary of calculations for trained network.

	1st training set	2nd training set	3rd training set	4th training set
	$x_1 = 1,$ $x_2 = 1,$ $y_d = 0$	$x_1 = 0,$ $x_2 = 1,$ $y_d = 1$	$x_1 = 1,$ $x_2 = 0,$ $y_d = 1$	$x_1 = 0,$ $x_2 = 0,$ $y_d = 0$
$w_{13}x_1 + w_{23}x_2 - \theta_3$	9.8741	3.5263	3.5257	-2.8221
$y_3 = \dfrac{1}{1 + e^{-(w_{13}x_1 + w_{23}x_2 - \theta_3)}}$	0.9999	0.9714	0.9714	0.0561
$w_{14}x_1 + w_{24}x_2 - \theta_4$	2.2340	-2.5613	-2.5614	-7.3566
$y_4 = \dfrac{1}{1 + e^{-(w_{14}x_1 + w_{24}x_2 - \theta_4)}}$	0.9033	0.0717	0.0717	0.0006
$w_{35}y_3 + w_{45}y_4 - \theta_5$	-4.1506	4.1807	4.1805	-4.0286
$y_5 = \dfrac{1}{1 + e^{-(w_{35}y_3 + w_{45}y_4 - \theta_5)}}$	0.0155	0.9849	0.9849	0.0175
y_d	0	1	1	0
$e_{k,p} = y_d - y_5$	-0.0155	0.0151	0.0151	-0.0175
$e_{k,p}^2 = (y_d - y_5)^2$	0.0002	0.0002	0.0002	0.0003

(UI), and micronaire value (MIC) were spun into three different yarn counts (16^s, 24^s, and 30^s Ne having twist multiplier of 3.6, 3.75, and 3.9, respectively) in ring spinning system. Table 2.5 shows a total of 51 data points comprising BS, UHML, UI, MIC, yarn English count (Ne), and yarn strength (cN/tex).

For the prediction of yarn strength using ANN, four fiber parameters such as BS, UHML, UI, and MIC along with yarn count (Ne) were considered as inputs. In the ANN model, the number of hidden layers is set to unity with six neurons. The tan-sigmoid and linear transfer functions are used as the activation functions for the hidden and output layers, respectively. The dataset was divided into training and testing data array using the k-fold cross-validation technique (Ghosh, 2010). In k-fold cross-validation, the initial dataset is randomly partitioned into k mutually exclusive subsets or folds $D_1, D_2, ..., D_k$, each of approximately equal size. The training and testing are performed k times. In iteration i, partition D_i is reserved as the test set and the remaining partitions are collectively used to train the model. In this method, each data point is used the same number of times for training and once for testing. Therefore, the validation of the model becomes more accurate and unbiased. The expected generalization accuracy was estimated as $\mu \pm \sigma$, where μ and σ are the mean and standard deviation of the percentage accuracies over k trials. A schematic diagram of k-fold cross-validation is illustrated in Fig. 2.20.

TABLE 2.5 Dataset of cotton fiber and yarn.

BS (cN/tex)	UHML (in.)	UI (%)	MIC (μg/in.)	Yarn count (Ne)	Yarn strength (cN/tex)
30.6	1.07	83.1	4.7	21.8	15.26
29	1.06	80.7	3.1	16.1	15.62
29	1.05	81.9	4.2	16.3	15.19
28.7	1.05	81	3.9	16.4	14.38
28.7	1.05	81	3.9	22.2	13.96
28.5	1.15	80.2	3.5	30	13.64
28.7	1.09	81	4.4	30.2	13.52
27.5	1.07	82.8	4.5	16.4	14.6
30.8	1.13	82.6	4.3	22.3	15.21
28.1	1.01	80.7	3.8	29.4	13.1
27.7	1.05	81.5	4.7	30.5	13.82
28.6	1.04	82.4	4.2	29.9	13.22
30.6	1.07	83.1	4.7	30.6	14.86

TABLE 2.5 Dataset of cotton fiber and yarn—cont'd

BS (cN/tex)	UHML (in.)	UI (%)	MIC (μg/in.)	Yarn count (Ne)	Yarn strength (cN/tex)
28.7	1.1	79.2	3.7	16.5	13.98
28.3	0.97	81.5	3.8	22.2	15.1
29	1.06	80.7	3.1	29.7	13.55
28.1	1.01	80.7	3.8	15.8	14.15
27.5	1.07	82.8	4.5	21.9	14.77
29	1.06	80.7	3.1	22.4	14.27
29.1	1.05	81.7	4	22.5	14.93
26.5	1.09	81.5	3.8	22.2	14.42
29.1	1.05	81.7	4	16.1	14.99
28.7	1.1	79.2	3.7	22.2	13.15
28.5	1.15	80.2	3.5	22.2	13.78
29.1	1.05	81.7	4	29.9	13.63
28.1	1.01	80.7	3.8	22	13.35
29.2	0.98	80	4.5	29.8	13.47
30.3	1.1	83.2	4.4	16.4	16.28
28.3	0.97	81.5	3.8	16.3	15.45
30.6	1.07	83.1	4.7	16.2	16.39
28.6	1.04	82.4	4.2	16.2	15.26
30.3	1.1	83.2	4.4	30.1	14.41
28.6	1.04	82.4	4.2	22	14.28
30.3	1.1	83.2	4.4	22.4	14.75
27.7	1.05	81.5	4.7	16.4	15.25
29.2	0.98	80	4.5	16	15.15
27.7	1.05	81.5	4.7	22	14.18
26.5	1.09	81.5	3.8	30.7	13.11
28.7	1.09	81	4.4	16.4	15.3
28.7	1.05	81	3.9	30.1	13.08
29	1.05	81.9	4.2	22.2	14.01
28.7	1.09	81	4.4	22.1	14.47

Continued

TABLE 2.5 Dataset of cotton fiber and yarn—cont'd

BS (cN/tex)	UHML (in.)	UI (%)	MIC (µg/in.)	Yarn count (Ne)	Yarn strength (cN/tex)
28.5	1.15	80.2	3.5	16.2	14.77
27.5	1.07	82.8	4.5	29.5	14.06
29	1.05	81.9	4.2	30	13.77
28.3	0.97	81.5	3.8	30.3	13.97
26.5	1.09	81.5	3.8	16	14.49
28.7	1.1	79.2	3.7	29.5	12.28
29.2	0.98	80	4.5	22.4	13.24
30.8	1.13	82.6	4.3	16.4	16.24
30.8	1.13	82.6	4.3	30.1	14.88

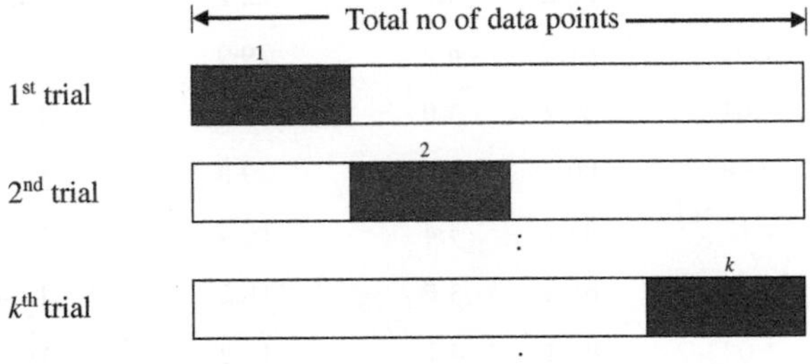

FIG. 2.20 Schematic representation of k-fold cross-validation.

The gradient descent backpropagation method is applied for training of the network. For each fold, the training parameters, i.e., the maximum number of iterations, learning rate (η), error tolerance limit, and minimum performance gradient tolerance limit are set to 40,000, 0.005, 0, and 10^{-5}, respectively. The mean absolute percentage accuracy (MAPA) between the actual and predicted values are used as performance criteria of the network. The MAPA is estimated using the following equation:

$$\text{MAPA} = \frac{100}{n} \left\{ \sum_{i=1}^{n} \left(1 - \left| \frac{e_i - p_i}{e_i} \right| \right) \right\} \qquad (2.26)$$

where e_i is the ith experimental value, p_i is the ith predicted value and n is the number of observations.

TABLE 2.6 Training and testing accuracies.

Fold no.	Training accuracy (%)	Testing accuracy (%)
1	98.85	97.17
2	99.05	98.53
3	98.93	96.91
4	98.81	97.42
5	98.91	96.78
6	98.95	95.95
7	98.90	98.59
8	98.90	95.66
9	98.83	92.28
10	98.62	98.26
Mean (μ)	98.87	96.76
Standard deviation (σ)	0.112	1.865

Ten-fold cross-validation was applied for the purpose of cross-validation. Accordingly, the ANN model was trained using 9 of the folds and tested on the fold left out. Thus, the training and testing were performed 10 times. A summary of MAPA obtained for training and testing sets in each fold is depicted in Table 2.6. The expected generalization accuracies were estimated as 98.87 ± 0.112 and 96.76 ± 1.865, respectively for training and testing sets. The learning accuracy on the training set was expectedly higher than the predictive accuracy on the test set due to the fact that the latter is performed on the unseen dataset. A MATLAB coding was used to execute the computational work.

2.5 MATLAB coding

2.5.1 MATLAB coding of ANN example given in Section 2.3

```
clc
clear
close all
x=[1 1
0 1
1 0
```

```
    0 0];
y=[0
   1
   1
   0];
W1=[0.1 0.2
    0.5 0.8];
IB=[-1 -1 -1];
B1=[0.4
    0.2];
W2=[1 -1];
B2=0.5;
alpha=0.1;
E=10^6;
j=0;k=10^6;
while E>0.0005
fori=1:length(y)
x1=x(i,:);
y1=y(i);
Ih=W1*x1'-B1;
Oh=logsig(Ih);
Iop=W2*Oh-B2;
Oop=logsig(Iop);
y_p=Oop;
e=y1-y_p;
delta_op=y_p.*(1-y_p).*e;
del_W2=(alpha*Oh.*delta_op)';
del_B2=alpha*IB(3)*delta_op;
delta_h=Oh.*(1-Oh).*delta_op.*W2';
del_W1=(alpha*x1'.*delta_h')';
del_B1=alpha*IB(1,1:2)'.*delta_h;
W1=W1+del_W1;
B1=B1+del_B1;
W2=W2+del_W2;
B2=B2+del_B2;
end
Ih=W1*x'-repmat(B1,1,length(y));
Oh =logsig(Ih);
Iop=W2*Oh-B2;
Oop=logsig(Iop);
y_p=Oop;
ee=y-y_p';
j=j+1;
```

```
E(j)=0.5*sum(ee.^2);
if j==k
break
end
end
j
W1
B1
W2
B2
y_p
```

2.5.2   MATLAB coding of ANN application given in Section 2.4

```
clc
close all
clear
format short g
data_tenacity = [30.6    1.07    83.1    4.7    21.8    15.26
                 29      1.06    80.7    3.1    16.1    15.62
                 29      1.05    81.9    4.2    16.3    15.19
                 28.7    1.05    81      3.9    16.4    14.38
                 28.7    1.05    81      3.9    22.2    13.96
                 28.5    1.15    80.2    3.5    30      13.64
                 28.7    1.09    81      4.4    30.2    13.52
                 27.5    1.07    82.8    4.5    16.4    14.6
                 30.8    1.13    82.6    4.3    22.3    15.21
                 28.1    1.01    80.7    3.8    29.4    13.1
                 27.7    1.05    81.5    4.7    30.5    13.82
                 28.6    1.04    82.4    4.2    29.9    13.22
                 30.6    1.07    83.1    4.7    30.6    14.86
                 28.7    1.1     79.2    3.7    16.5    13.98
                 28.3    0.97    81.5    3.8    22.2    15.1
                 29      1.06    80.7    3.1    29.7    13.55
                 28.1    1.01    80.7    3.8    15.8    14.15
                 27.5    1.07    82.8    4.5    21.9    14.77
                 29      1.06    80.7    3.1    22.4    14.27
                 29.1    1.05    81.7    4      22.5    14.93
                 26.5    1.09    81.5    3.8    22.2    14.42
                 29.1    1.05    81.7    4      16.1    14.99
                 28.7    1.1     79.2    3.7    22.2    13.15
                 28.5    1.15    80.2    3.5    22.2    13.78
                 29.1    1.05    81.7    4      29.9    13.63
                 28.1    1.01    80.7    3.8    22      13.35
```

```
                29.2    0.98    80      4.5     29.8    13.47
                30.3    1.1     83.2    4.4     16.4    16.28
                28.3    0.97    81.5    3.8     16.3    15.45
                30.6    1.07    83.1    4.7     16.2    16.39
                28.6    1.04    82.4    4.2     16.2    15.26
                30.3    1.1     83.2    4.4     30.1    14.41
                28.6    1.04    82.4    4.2     22      14.28
                30.3    1.1     83.2    4.4     22.4    14.75
                27.7    1.05    81.5    4.7     16.4    15.25
                29.2    0.98    80      4.5     16      15.15
                27.7    1.05    81.5    4.7     22      14.18
                26.5    1.09    81.5    3.8     30.7    13.11
                28.7    1.09    81      4.4     16.4    15.3
                28.7    1.05    81      3.9     30.1    13.08
                29      1.05    81.9    4.2     22.2    14.01
                28.7    1.09    81      4.4     22.1    14.47
                28.5    1.15    80.2    3.5     16.2    14.77
                27.5    1.07    82.8    4.5     29.5    14.06
                29      1.05    81.9    4.2     30      13.77
                28.3    0.97    81.5    3.8     30.3    13.97
                26.5    1.09    81.5    3.8     16      14.49
                28.7    1.1     79.2    3.7     29.5    12.28
                29.2    0.98    80      4.5     22.4    13.24
                30.8    1.13    82.6    4.3     16.4    16.24
                30.8    1.13    82.6    4.3     30.1    14.88];
[s1,s2]=size(data_tenacity);
random_sample=randperm(s1);%Data randomisation
data_tenacity=data_tenacity(random_sample,:);
%Input data normalisation
for j=1:5
    norm_data(:,j)=(data_tenacity(:,j)-min(data_tenacity(:,j)))/
    (max(data_tenacity(:,j))-min(data_tenacity(:,j)));
    end
n=s1;%Number of experiments or data points
input_array=norm_data;
target_array=data_tenacity(:,end);
%k-fold cross validation
% Division of training and testing arrays of input pattern
k=10;% for 10-fold cross validation
n=s1;%Number of experiments or data points
sizeG=round(n/k);%groupsize
Random_sample=randperm(n);%Data randomisation
%Creating K folds
```

```
j=1;
fori=1:k
ifi<k
        folds(i,:)=Random_sample(j:i*sizeG);
        j=1+i*sizeG;
elseifi==k
lastfold=Random_sample((1+(i-1)*sizeG):end);
end
end
fori=1:k
ifi~=k
input_array_testing=input_array(folds(i,:),:);
target_array_testing=target_array(folds(i,:),:);
elseifi==k
input_array_testing=input_array(lastfold,:);
target_array_testing=target_array(lastfold,:);
end
ifi==1
        input_array_training1=input_array(folds((i+1):end,:),:);
        input_array_training2=input_array(lastfold,:);
        target_array_training1=target_array(folds((i+1):
end,:),:);
        target_array_training2=target_array(lastfold,:);
        input_array_training=[input_array_training1;
input_array_training2];
target_array_training=[target_array_training1;
target_array_training2];
elseifi==k
input_array_training=input_array(folds(1:end,:),:);
target_array_training=target_array(folds(1:end,:),:);
else
        input_array_training1=input_array(folds([1:(i-1)   (i+1):
end],:),:);
        input_array_training2=input_array(lastfold,:);
        input_array_training=[input_array_training1;
input_array_training2];
        target_array_training1=target_array(folds([1:(i-1) (i+1):
end],:),:);
        target_array_training2=target_array(lastfold,:);
        target_array_training=[target_array_training1;
target_array_training2];
end
% ANN model
```

```
net= newff(input_array_training', target_array_training', 6, {'tan-
    sig', 'purelin'}, 'traingd');
net.trainParam.show= 1000;
net.trainParam.lr= 0.005;
net.trainParam.epochs= 40000;
net.trainParam.goal=0;
net.trainParam.goal= 0.0001;
net.trainParam.min_grad=1e-5;
net.divideParam.testRatio = 0;
net.divideParam.valRatio = 0;
%
net =train(net, input_array_training', target_array_training');
predicted_value_testing=sim(net,input_array_testing');
predicted_value_training=sim(net,input_array_training');
%Prediction statistics
    y=target_array_testing;
    y1=predicted_value_testing;
mean_accuracy_testing(i)=mean(100-((abs(y-y1')./y)*100));
    z=target_array_training;
    z1=predicted_value_training;
mean_accuracy_training(i)=mean(100-((abs(z-z1')./z)*100));
end
%Prediction results
Testing_accuracy=mean_accuracy_testing'
Training_accuracy=mean_accuracy_training'
grand_mean_accuracy_testing=mean(Testing_accuracy)
grand_mean_accuracy_training=mean(Training_accuracy)
S1=std(Testing_accuracy)
S2=std(Training_accuracy)
```

2.6   Summary

The chapter on ANN presents an overview of this powerful machine learning algorithm inspired by the structure and function of the human brain. It explores the basic concepts, architecture, and training methods associated with ANN. The chapter begins by introducing the basic building blocks of an ANN, called neurons or nodes which are interconnected and act collectively to process and communicate the input signals. Various types of activation functions which determine how the neuron responds to input signals have been discussed. This chapter also touches upon the concept of weights and biases, which are adjustable parameters that determine the strength of connections between neurons. A feedforward backpropagation ANN is discussed, where information flows in the forward direction, whereas the error signal propagates in the backward direction and thereby adjusting the network's weights and biases. The iterative nature of training and the concept of gradient descent for optimizing the network performance are explained. It then discusses a worked-out example showing the step-by-step procedure of a feedforward backpropagation ANN model. It also includes the results of a study on the application of ANN in predicting yarn strength from fiber and yarn properties. This chapter serves as a strong foundation for understanding and exploring the potential of ANN in solving complex problems across different domains.

References

Ghosh, A. (2010). Forecasting of cotton yarn properties using intelligent machines. *Research Journal of Textile and Apparel, 14*(3), 55–61.

Haykin, S. (2004). *Neural networks: A comprehensive foundation* (2nd ed., pp. 161–175). Singapore: Pearson Education.

Kartalopoulos, S. V. (2000). *Understanding neural networks and fuzzy logic: Basic concepts and applications* (pp. 75–82). New Delhi: Prentice-Hall of India Pvt. Ltd.

Khanna, T. (1990). *Foundations of neural networks.* Addison-Wesley Longman Publishing Co., Inc.

McCulloch, W. S., & Pitts, W. (1943). A logical calculus of the ideas immanent in nervous activity. *The Bulletin of Mathematical Biophysics, 5*, 115–133.

Rosenblatt, F. (1958). The perceptron: A probabilistic model for information storage and organization in the brain. *Psychological Review, 65*(6), 386–408.

Yegnanarayana, B. (2009). *Artificial neural networks.* PHI Learning Pvt. Ltd.

Zurada, J. M. (2003). *Introduction to artificial neural systems.* Mumbai: Jaico Publishing House.

Chapter 3

Support vector machines

3.1 Introduction

Support vector machine (SVM) is a promising machine learning method that belongs to the class of supervised learning algorithms. It has gained significant popularity in the classification of both linear and nonlinear data. SVM is originally designed for binary classification, nevertheless, it can be extended to handle multi-class classification. Moreover, it is very much useful for regression tasks. SVM uses kernel trick which transforms the original training data into a higher dimension feature space by a nonlinear mapping. Kernel functions such as polynomial and Gaussian are used for the nonlinear mapping. In the higher dimension feature space, it searches for linear optimal separating hyperplane or decision boundary. The optimal separating hyperplane is determined by selecting a subset of training samples, called support vectors, which lie closest to the decision boundary. The ability of SVM to handle nonlinear data through the kernel trick and the emphasis on maximizing the margin makes it a powerful tool of machine learning. Training of an SVM involves solving a convex optimization problem, typically formulated as a quadratic programming task. The history of SVM dates back to the early 1960s, with the work by Vapnik and Chervonenkis (1964) on statistical learning theory. However, it gained widespread attention only during the 1990s, with the pioneering work by Vapnik (1995). Ever since, SVM has evolved (Fung & Mangasarian, 2005; Suykens et al., 2002) and become a centerpiece of machine learning, finding applications in numerous fields, such as engineering, bioinformatics, finance, image recognition, natural language processing, and so on so forth. This chapter delves a comprehensive introduction to support vector classification and support vector regression, their basic principles in steps with illustrative examples and applications in the textile domain.

3.2 Support vector classification (SVC)

SVM as a classifier separates data points linearly in high-dimension feature space by maximizing the distance between different classes. In SVC, the separation is done by an algorithm based on a convex optimization problem where the minimization of a quadratic function under linear inequality constraints is done from optimization theory. As real-world problems are often very complex, implicit mapping of data points through kernel functions to effect linearity in

Artificial Intelligence in Textile Engineering. https://doi.org/10.1016/B978-0-443-15395-2.00001-6

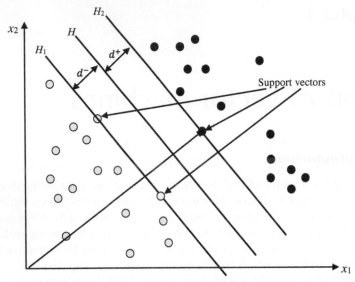

FIG. 3.1 Optimal separating hyperplane with support vectors.

high-dimension feature space is quite a necessity. The concept of SVC is discussed later (Cristianini & Shawe-Taylor, 2000; Hamel, 2009; Schölkopf & Smola, 2002; Vapnik, 1995).

Consider the problem of separating the set of training vectors belonging to two separate classes, (x_i, y_i); $x_i \in \Re^n$; $y_i \in \{-1, +1\}$; $i = 1, 2, ..., N$ as depicted in Fig. 3.1. Theoretically, an infinity number of hyperplanes in $\Re^n$ which are parameterized by w and a constant b can be conceived that can separate the data into two classes. Our objective is to find a hyperplane $f(x) = sign(w^T x + b)$ that correctly classifies the data. The optimal hyperplane H should be such that

$$w^T x_i + b \geq +1, \quad \text{when } y_i = +1 \tag{3.1}$$

and

$$w^T x_i + b \leq -1, \quad \text{when } y_i = -1 \tag{3.2}$$

This corresponds to the optimal separating hyperplane

$$H : w^{*T} x_i + b^* = 0 \tag{3.3}$$

The corresponding margins are defined as

$$H_1 : w^{*T} x_i + b^* = +1 \tag{3.4}$$

and

$$H_2 : w^{*T} x_i + b^* = -1 \tag{3.5}$$

where w^* and b^* are the optimal solutions.

It is desirable to have a classifier with as big a margin as possible for optimal separation of data points. The distance between H and H_1 is given by $\frac{1}{\|w\|}$. Therefore, the total margin, which is the distance between H_1 and H_2 is $\frac{2}{\|w\|}$ the details of which are furnished in the following theorem.

Theorem 3.1 *The distance between two margins is* $\frac{2}{\|w\|}$.

Proof *Suppose a given data can be classified into two classes as shown in Fig. 3.2. Mathematically, we distinguish a particular observation x^+ as belonging to class I if*

$$w \cdot x^+ + b \geq +1 \tag{3.6}$$

Or, an observation x^- belonging to class II if

$$w \cdot x^- + b \leq -1 \tag{3.7}$$

For any observation x^+ belonging to the upper margin, we have

$$w \cdot x^+ + b = 1 \tag{3.8}$$

and similarly, for any observation x^- belonging to the lower margin, we can write

$$w \cdot x^- + b = -1 \tag{3.9}$$

Geometrically, w is the direction ratios of the lines perpendicular to the hyperplane given in (3.9) and the perpendicular distance between the two

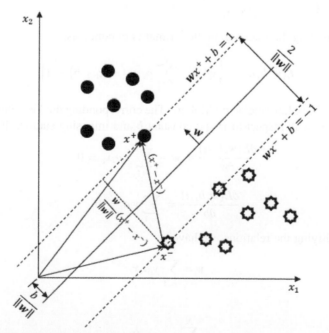

FIG. 3.2 The distance between the margins.

margins is the component or projection of the vector $x^+ - x^-$ along the direction w (see Fig. 3.2). Note that the unit vector along the direction of w is $\dfrac{w}{\|w\|}$. Thus the width (d) between the two lines can be obtained as

$$\frac{w}{\|w\|} \cdot x^+ - \frac{w}{\|w\|} \cdot x^- = \frac{1}{\|w\|}(w \cdot x^+ - w \cdot x^-) \tag{3.10}$$

Putting the value of w·x^+ and w·x^- from (3.8) and (3.9) into (3.10) we get

$$d = \frac{2}{\|w\|}$$

This completes the proof.

To maximize the margin d, we need to minimize $\|w\|$. The conditions that there are no data points between H_1 and H_2 are given by Eqs. (3.1) and (3.2) which can be combined as

$$y_i(w^T x_i + b) \geq 1 \tag{3.11}$$

The maximum margin classifier can be obtained by solving the following optimization problem:

$$\underset{w,b}{\text{Minimize}}\, \frac{1}{2} w^T w \tag{3.12}$$

Subject to

$$y_i(w^T x_i + b) \geq 1, \quad i = 1, 2, ..., N.$$

Formulating the Lagrangian, the primal is obtained as,

$$L(w, b, \alpha) = \frac{1}{2} w^T w - \sum_{i=1}^{N} \alpha_i [y_i(w^T x_i + b) - 1] \tag{3.13}$$

where $\alpha_i \geq 0$ are Lagrangian multipliers. The corresponding dual is found by differentiating with respect to w and constant b and imposing stationarity,

$$\frac{\partial L(w, b, \alpha)}{\partial w} = w - \sum_{i=1}^{N} y_i \alpha_i x_i = 0 \tag{3.14}$$

$$\frac{\partial L(w, b, \alpha)}{\partial b} = \sum_{i=1}^{N} y_i \alpha_i = 0 \tag{3.15}$$

Simplifying the relations, we have

$$w = \sum_{i=1}^{N} y_i \alpha_i x_i \tag{3.16}$$

and

$$\sum_{i=1}^{N} y_i \alpha_i = 0 \tag{3.17}$$

Resubstituting back to the primal Lagrangian we get

$$L(w, b, \alpha) = \frac{1}{2}w^T w - \sum_{i=1}^{N} \alpha_i \left[y_i \left(w^T x_i + b \right) - 1 \right]$$

$$= \frac{1}{2} \sum_{i=1}^{N} \sum_{j=1}^{N} y_i y_j \alpha_i \alpha_j x_i^T x_j - \sum_{i=1}^{N} \sum_{j=1}^{N} y_i y_j \alpha_i \alpha_j x_i^T x_j + \sum_{i=1}^{N} \alpha_i \qquad (3.18)$$

$$= \sum_{i=1}^{N} \alpha_i - \frac{1}{2} \sum_{i=1}^{N} \sum_{j=1}^{N} y_i y_j \alpha_i \alpha_j x_i^T x_j$$

Hence the dual problem becomes
Maximize

$$Q(\alpha) = \sum_{i=1}^{N} \alpha_i - \frac{1}{2} \sum_{i=1}^{N} \sum_{j=1}^{N} y_i y_j \alpha_i \alpha_j x_i^T x_j \qquad (3.19)$$

Subject to

$$\sum_{i=1}^{N} y_i \alpha_i = 0$$

$$\alpha_i \geq 0, \quad i = 1, 2, ..., N.$$

The dual problem can be easily solved by readily available quadratic programming solvers to give α^*. The weight vector that realizes the maximal margin hyperplane is given by

$$w^* = \sum_{i=1}^{N} y_i \alpha_i^* x_i \qquad (3.20)$$

The value of b does not appear in the dual problem and so b^* is found from primal constraints

$$b^* = -\frac{\max\limits_{y_i = -1} \left(w^{*T} x_i \right) + \min\limits_{y_i = +1} \left(w^{*T} x_i \right)}{2} \qquad (3.21)$$

The corresponding Karush-Kuhn-Tucker (KKT) complementarity condition is given by

$$\alpha_i^* \left[y_i \left(w^{*T} x_i + b^* \right) - 1 \right] = 0, \quad i = 1, 2, ..., N \qquad (3.22)$$

where α^*, w^*, and b^* are the optimal solutions.

The interesting part of the above relation is that, in order to α_i^* to be non-zero, $[y_i(w^{*T}x_i + b^*) - 1] = 0$ and all other α_i^*'s are 0. This shows that, only those x_i's are important which lies on the hyperplanes given by $y_i(w^{*T}x_i + b^*) - 1 = 0$. That is why these vectors are called support vectors (sv). This also shows that the points which are not support vectors have no effect on the solution. With the help of this idea, we can state the following theorem.

Theorem 3.2 *Let $\{(x_1, y_1), (x_2, y_2), \ldots, (x_n, y_n)\}$ be a linearly separable training sample and α^*, b^* is the solution of the dual optimization problem (3.19). Then the weight vector w attains its minima corresponding to the maximal margin hyperplane with geometric mean*

$$\frac{1}{\|w\|} = \left(\sum_{i \in sv} \alpha_i^* \right)^{-\frac{1}{2}}.$$

Proof *For $i \in sv$ where sv is the set of indices corresponding to the support vectors,*

$$y_j \left[\left(\sum_{i \in sv} y_i \alpha_i^* x_i x_j + b^* \right) \right] = 1.$$

Therefore,

$$\|w\|^2 = \langle w, w \rangle = \sum_{i, j \in sv} y_i y_j \alpha_i^* \alpha_j^* \langle x_i, x_j \rangle$$

$$= \sum_{j \in sv} \alpha_j^* y_j \sum_{i \in sv} y_i \alpha_i^* \langle x_i, x_j \rangle$$

$$= \sum_{j \in sv} \alpha_j^* \left(1 - y_j b^* \right) \tag{3.23}$$

$$= \sum_{j \in sv} \alpha_j^* \quad \left[\because \sum_{j \in sv} \alpha_j^* y_j = 0 \right]$$

This completes the proof.

In the dual solution, it turns out that most of the α^* are zero. Non-zero values occur for the points that are closest to the hyperplane. These are known as the support vectors (sv). The optimal hyperplane is given by

$$f(x) = \sum_{i=1}^{N} y_i \alpha_i^* x_i^T x_j + b^*$$

$$= \sum_{i \in sv} y_i \alpha_i^* x_i^T x_j + b^* \tag{3.24}$$

In the case where a linear boundary is inappropriate, the SVC can map the input vector x_i into a high-dimensional feature space. Among acceptable mappings are polynomials, Gaussian radial basis functions, and certain sigmoid functions. In the case of nonlinear mapping into a high-dimensional feature space, the optimization problem becomes

Maximize

$$Q(\alpha) = \sum_{i=1}^{N} \alpha_i - \frac{1}{2} \sum_{i=1}^{N} \sum_{j=1}^{N} y_i y_j \alpha_i \alpha_j K(x_i, x_j) \tag{3.25}$$

Subject to

$$\sum_{i=1}^{N} y_i \alpha_i = 0$$

$$\alpha_i \geq 0, i = 1, 2, \dots, N.$$

The corresponding optimal hyperplane is given by

$$f(x) = \sum_{i=1}^{N} y_i \alpha_i^* K(x_i, x_j) + b^*$$

$$= \sum_{i \in sv} y_i \alpha_i^* K(x_i, x_j) + b^* \tag{3.26}$$

The function $K(x_i, x_j)$ is a kernel function that projects the data into a high-dimensional feature space and thereby increases the computational power of the linear learning machine. A kernel may be defined as a function K, such that for all $x_i, x_j \in X$

$$K(x_i, x_j) = \phi^T(x_i)\phi(x_j) \tag{3.27}$$

where ϕ is a mapping from input space X to an inner product feature space F.

The use of kernels makes it possible to map the data implicitly into a feature space and to train a linear machine in such a space, potentially side-stepping the computational problems inherent in evaluating the feature map. Therefore, we need to first create a complicated feature space, then work out what the inner product in that space would be, and finally find a direct method of computing that value in terms of the original inputs. Different types of kernel functions are exemplified in Table 3.1. Fig. 3.3 illustrates an example of a feature mapping, where linear classification of data cannot be possible in the input space; however, it can be possible in the feature space.

The main problem with the maximal margin classifier is that it always produces a perfectly consistent hypothesis, which is a hypothesis with no training

TABLE 3.1 Different types of kernel functions.

Type of function	Expression
Linear	$K(x_i, x_j) = x_i^T x_j$
Polynomial	$K(x_i, x_j) = (x_i^T x_j + t)^d$
Gaussian radial basis	$K(x_i, x_j) = \exp\left(-\dfrac{\|x_i - x_j\|^2}{2\sigma^2}\right)$
Exponential radial basis	$K(x_i, x_j) = \exp\left(-\dfrac{\|x_i - x_j\|}{2\sigma^2}\right)$
Multilayer perceptron	$K(x_i, x_j) = \tanh(sx_i^T x_j + t^2)$

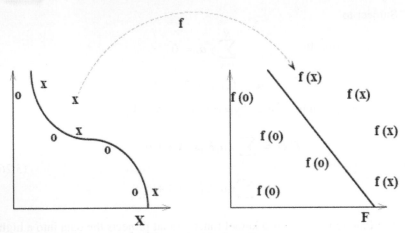

FIG. 3.3 A feature map can simplify the classification tasks.

error. In real data, where noise can always be present, this can result in a brittle estimator. These problems can be overcome by using the soft-margin optimization, where we need to introduce slack variables (ξ_i) to allow the margin constraints to be violated subject to

$$y_i\left(w^T x_i + b\right) \geq 1 - \xi_i \tag{3.28}$$

$$\xi_i \geq 0, i = 1, 2, ..., l.$$

Thus, the soft-margin optimization problem becomes:

$$\underset{\xi,\, w\, b}{\text{Minimize}}\ \frac{1}{2} w^T w + C \sum_{i=1}^{l} \xi_i \ (\text{for } 1 - \text{norm soft margin}) \tag{3.29}$$

Or,

$$\underset{\xi,\, w\, b}{\text{Minimize}}\ \frac{1}{2} w^T w + C \sum_{i=1}^{l} \xi_i^2 \ (\text{for } 2 - \text{norm soft margin}) \tag{3.30}$$

Subject to

$$y_i\left(w^T x_i + b\right) \geq 1 - \xi_i, \quad i = 1, 2, ..., l.$$

where C is a prespecified value.

3.2.1 Step-by-step working principle of a linearly separable SVC

To illustrate the step-by-step procedure for the design of a support vector machine, we first consider the following linearly separable binary classification problem as shown in Table 3.2. The scatter plots of two classes (1 and -1) are illustrated in Fig. 3.4.

TABLE 3.2 Linearly separable binary classification problem.

Input vector		Response
x_1	x_2	y
0.5	0.5	1
0.3	0.7	1
0.1	0.1	−1
0.2	0	−1

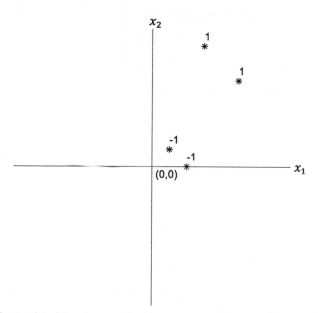

FIG. 3.4 Scatter plot of linearly separable data.

In this classification problem, we have input vector x with two dimensions as follows:

$$x = [x_1 \; x_2]^T$$

To find the dot product, we express

$$x_i = [x_{i1} \; x_{i2}]^T = \begin{bmatrix} 0.5 & 0.3 & 0.1 & 0.2 \\ 0.5 & 0.7 & 0.1 & 0 \end{bmatrix}$$

$$x_j = [x_{j1} \; x_{j2}]^T = \begin{bmatrix} 0.5 & 0.3 & 0.1 & 0.2 \\ 0.5 & 0.7 & 0.1 & 0 \end{bmatrix}$$

Thus, the dot product of vectors x_i and x_j can be written as

$$x_i^T x_j = [x_{i1} \; x_{i2}] \begin{bmatrix} x_{j1} \\ x_{j2} \end{bmatrix}$$

$$= \begin{bmatrix} 0.5 & 0.5 \\ 0.3 & 0.7 \\ 0.1 & 0.1 \\ 0.2 & 0 \end{bmatrix} \times \begin{bmatrix} 0.5 & 0.3 & 0.1 & 0.2 \\ 0.5 & 0.7 & 0.1 & 0 \end{bmatrix}$$

$$= \begin{bmatrix} 0.5 & 0.5 & 0.1 & 0.1 \\ 0.5 & 0.58 & 0.1 & 0.06 \\ 0.1 & 0.1 & 0.02 & 0.02 \\ 0.1 & 0.06 & 0.02 & 0.04 \end{bmatrix}$$

The optimization problem in the dual form is given by
Maximize

$$Q(\alpha) = \sum_{i=1}^{N} \alpha_i - \frac{1}{2} \sum_{i=1}^{N} \sum_{j=1}^{N} \alpha_i \alpha_j y_i y_j x_i^T x_j$$

Subject to

$$\sum_{i=1}^{N} y_i \alpha_i = 0$$

$$\alpha_i \geq 0, \quad i = 1, 2, \ldots, N.$$

By substituting the values of $x_i^T x_j$ and response variable y_i, the optimization problem in the dual form becomes
Maximize

$$\alpha_1 + \alpha_2 + \alpha_3 + \alpha_4 - \frac{1}{2} \Big(0.5\alpha_1^2 + 0.58\alpha_2^2 + 0.02\alpha_3^2 + 0.04\alpha_4^2 + \alpha_1\alpha_2$$

$$- 0.2\alpha_1\alpha_3 - 0.2\alpha_1\alpha_4 - 0.2\alpha_2\alpha_3 - 0.12\alpha_2\alpha_4 + 0.04\alpha_3\alpha_4 \Big)$$

Subject to

$$\alpha_1 + \alpha_2 - \alpha_3 - \alpha_4 = 0$$
$$\alpha_i \geq 0, \quad i = 1, 2, \ldots, N.$$

The Lagrangian of this optimization problem is

$$L(\alpha, \lambda) = \alpha_1 + \alpha_2 + \alpha_3 + \alpha_4 - \frac{1}{2}\left(0.5\alpha_1^2 + 0.58\alpha_2^2 + 0.02\alpha_3^2 + 0.04\alpha_4^2\right.$$
$$+ \alpha_1\alpha_2 - 0.2\alpha_1\alpha_3 - 0.2\alpha_1\alpha_4 - 0.2\alpha_2\alpha_3 - 0.12\alpha_2\alpha_4$$
$$\left. + 0.04\alpha_3\alpha_4\right) + \lambda(\alpha_1 + \alpha_2 - \alpha_3 - \alpha_4)$$

Now,

$$\frac{\partial L}{\partial \alpha_1} = 0 \Rightarrow -0.5\alpha_1 - 0.5\alpha_2 + 0.1\alpha_3 + 0.1\alpha_4 + \lambda = -1$$

$$\frac{\partial L}{\partial \alpha_2} = 0 \Rightarrow -0.5\alpha_1 - 0.58\alpha_2 + 0.1\alpha_3 + 0.06\alpha_4 + \lambda = -1$$

$$\frac{\partial L}{\partial \alpha_3} = 0 \Rightarrow 0.1\alpha_1 + 0.1\alpha_2 - 0.02\alpha_3 - 0.02\alpha_4 - \lambda = -1$$

$$\frac{\partial L}{\partial \alpha_4} = 0 \Rightarrow 0.1\alpha_1 + 0.06\alpha_2 - 0.02\alpha_3 - 0.04\alpha_4 - \lambda = -1$$

$$\frac{\partial L}{\partial \lambda} = 0 \Rightarrow \alpha_1 + \alpha_2 - \alpha_3 - \alpha_4 + 0. \quad \lambda = 0$$

This system of equations can be written as

$$A\alpha = B$$

where

$$A = \begin{bmatrix} -0.5 & -0.5 & 0.1 & 0.1 & 1 \\ -0.5 & -0.58 & 0.1 & 0.06 & 1 \\ 0.1 & 0.1 & -0.02 & -0.02 & -1 \\ 0.1 & 0.06 & -0.02 & -0.04 & -1 \\ 1 & 1 & -1 & -1 & 0 \end{bmatrix}$$

$$B = [-1 \quad -1 \quad -1 \quad -10]^T$$

$$\alpha = [\alpha_1 \quad \alpha_2 \quad \alpha_3 \quad \alpha_4 \quad \lambda]^T$$

Solving by matrix inversion method, we get

$$\alpha_1^* = 6.2498$$
$$\alpha_2^* = 0.0002$$
$$\alpha_3^* = 6.2497$$
$$\alpha_4^* = 0.0003$$

Hence, the optimum value of the objective function $Q(\alpha)$ of the dual problem becomes

$$Q\left(\alpha^*\right) = 6.2498 + 0.0002 + 6.2497 + 0.0003 - \frac{1}{2}\Big\{0.5 \times 6.2498^2 + 0.58$$

$$\times 0.0002^2 + 0.02 \times 6.2497^2 + 0.04 \times 0.0003^2 + 6.2498 \times 0.0002$$

$$- 0.2 \times 6.2498 \times 6.2497 - 0.2 \times 6.2498 \times 0.0002 - 0.2 \times 0.0002$$

$$\times 6.2497 - 0.12 \times 0.0002 \times 0.0003 + 0.04 \times 6.2497 \times 0.0003\Big\}$$

$$= 6.25$$

Also, we get

$$\sum_{i=1}^{N} \alpha_i^* = 6.2498 + 0.0002 + 6.2497 + 0.0003$$

$$= 12.5$$

From Eq. (3.20), we have

$$w^* = \sum_{i=1}^{N} y_i \alpha_i^* x_i$$

Hence,

$$w^{*T} w^* = \sum_{i=1}^{N} \sum_{j=1}^{N} \alpha_i^* \alpha_j^* y_i y_j x_i^T x_j$$

Therefore, we can write

$$Q\left(\alpha^*\right) = \sum_{i=1}^{N} \alpha_i^* - \frac{1}{2} \sum_{i=1}^{N} \sum_{j=1}^{N} \alpha_i^* \alpha_j^* y_i y_j x_i^T x_j$$

$$= \sum_{i=1}^{N} \alpha_i^* - \frac{1}{2} w^{*T} w^*$$

$$\text{Or,}\ \frac{1}{2} w^{*T} w^* = \sum_{i=1}^{N} \alpha_i^* - Q\left(\alpha^*\right)$$

$$\text{Or,}\ \frac{1}{2} w^{*T} w^* = 12.5 - 6.25 = 6.25$$

$$\text{Or,}\ w^{*T} w^* = 12.5$$

Hence,

$$\left\| w^* \right\| = \sqrt{12.5} = 3.54$$

Hence the optimum weight vector is

$$w^* = \sum_{i=1}^{N} y_i \alpha_i^* x_i$$

$$= \alpha_1^* x_1 + \alpha_2^* x_2 - \alpha_3^* x_3 - \alpha_4^* x_4$$

$$= 6.2498 \begin{bmatrix} 0.5 \\ 0.5 \end{bmatrix} + 0.0002 \begin{bmatrix} 0.3 \\ 0.7 \end{bmatrix} - 6.2497 \begin{bmatrix} 0.1 \\ 0.1 \end{bmatrix} - 0.0003 \begin{bmatrix} 0.2 \\ 0 \end{bmatrix}$$

$$= \begin{bmatrix} 3.1249 \\ 3.1249 \end{bmatrix} + \begin{bmatrix} 0.00006 \\ 0.00014 \end{bmatrix} - \begin{bmatrix} 0.62497 \\ 0.62497 \end{bmatrix} - \begin{bmatrix} 0.00006 \\ 0 \end{bmatrix}$$

$$= \begin{bmatrix} 2.5 \\ 2.5 \end{bmatrix}$$

Thus, we have

$$w^{*T} x_i = [2.5 \ 2.5] \times \begin{bmatrix} 0.5 & 0.3 & 0.1 & 0.2 \\ 0.5 & 0.7 & 0.1 & 0 \end{bmatrix}$$

$$= [2.5 \ 2.5 \ 0.5 \ 0.5]$$

Obviously,

$$\min\left(w^{*T} x_i\right) = 2.5, \quad \text{when } y_i = +1$$

and

$$\max\left(w^{*T} x_i\right) = 0.5, \quad \text{when } y_i = -1$$

Hence, from Eq. (3.21), the optimum value of b is

$$b^* = -\frac{\displaystyle\max_{y_i=-1}\left(w^{*T} x_i\right) + \min_{y_i=+1}\left(w^{*T} x_i\right)}{2}$$

$$= -\frac{0.5 + 2.5}{2}$$

$$= -1.5$$

Therefore, the equation of optimal separating hyperplane is

$$w^{*T} x_i + b^* = 0$$

$$\text{Or, } [2.5 \ 2.5] \begin{bmatrix} x_1 \\ x_2 \end{bmatrix} - 1.5 = 0$$

$$\text{Or, } 2.5x_1 + 2.5x_2 - 1.5 = 0$$

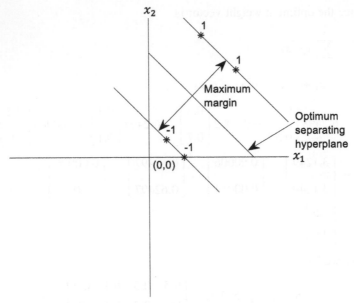

FIG. 3.5 Optimum separating hyperplane and margins.

Also, the equation of the first margin is

$$w^{*T} x_i + b^* = +1$$

$$\text{Or}, 2.5x_1 + 2.5x_2 - 1.5 = 1$$

Similarly, the equation of the second margin is

$$w^{*T} x_i + b^* = -1$$

$$\text{Or}, 2.5x_1 + 2.5x_2 - 1.5 = -1$$

Fig. 3.5 depicts the optimum separating hyperplane with two margins. Here, all four data points are support vectors as all of them are lying on the margins.

3.2.2 Step-by-step working principle of a nonlinearly separable SVC

To illustrate the step-by-step procedure for the design of a support vector machine, we consider the following nonlinearly separable binary classification problem as shown in Table 3.3. The scatter plots of two classes (1 and -1) are illustrated in Fig. 3.6.

We have,

$$x = [x_1 \quad x_2]^T$$
$$x_i = [x_{i1} \quad x_{i2}]^T$$
$$x_j = [x_{j1} \quad x_{j2}]^T$$

TABLE 3.3 Nonlinearly separable binary classification problem.

Input vector		Response
x_1	x_2	y
1	1	1
1	−1	1
−1	−1	1
−1	1	1
0.5	0.5	−1
0.5	−0.5	−1
−0.5	−0.5	−1
−0.5	0.5	−1

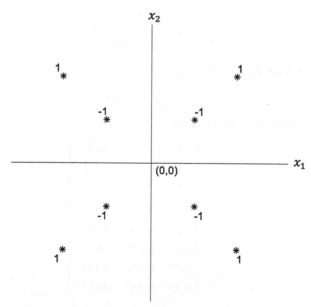

FIG. 3.6 Scatter plot of nonlinearly separable data.

Thus, the dot product of vectors x_i and x_j can be written as

$$x_i^T x_j = \begin{bmatrix} x_{i1} & x_{i2} \end{bmatrix} \begin{bmatrix} x_{j1} \\ x_{j2} \end{bmatrix}$$

$$= x_{i1} x_{j1} + x_{i2} x_{j2}$$

(3.31)

By taking square on both sides of Eq. (3.31), we can write

$$\left(x_i^T x_j\right)^2 = \left(x_{i1}x_{j1} + x_{i2}x_{j2}\right)^2$$
$$= x_{i1}^2 x_{j1}^2 + x_{i2}^2 x_{j2}^2 + 2x_{i1}x_{i2}x_{j1}x_{j2}$$

We assume that the inner product kernel $K(x_i, x_j)$ has the following homogeneous polynomial form

$$K\left(x_i, x_j\right) = \left(x_i^T x_j\right)^2$$
$$= x_{i1}^2 x_{j1}^2 + x_{i2}^2 x_{j2}^2 + 2x_{i1}x_{i2}x_{j1}x_{j2} \qquad (3.32)$$
$$= \left[x_{i1}^2 x_{i2}^2 \sqrt{2}x_{i1}x_{i2}\right] \begin{bmatrix} x_{j1}^2 \\ x_{j2}^2 \\ \sqrt{2}x_{j1}x_{j2} \end{bmatrix}$$

Hence, the mapping function which maps the input vectors x_i in the linearly separable highdimensional feature space can be expressed as

$$\phi(x_i) = \left[x_{i1}^2 x_{i2}^2 \sqrt{2}x_{i1}x_{i2}\right]^T$$

Similarly,

$$\phi(x_j) = \left[x_{j1}^2 x_{j2}^2 \sqrt{2}x_{j1}x_{j2}\right]^T$$

Therefore, from Eq. (3.27) we get

$$K\left(x_i, x_j\right) = \phi^T\left(x_i\right)\phi\left(x_j\right)$$

From the input data, we find that

$$\phi^T(x_i) = \begin{bmatrix} 1 & 1 & \sqrt{2} \\ 1 & 1 & -\sqrt{2} \\ 1 & 1 & \sqrt{2} \\ 1 & 1 & -\sqrt{2} \\ 0.25 & 0.25 & 0.354 \\ 0.25 & 0.25 & -0.354 \\ 0.25 & 0.25 & 0.354 \\ 0.25 & 0.25 & -0.354 \end{bmatrix}$$

and

$$\phi(x_j) = \begin{bmatrix} 1 & 1 & 1 & 1 & 0.25 & 0.25 & 0.25 & 0.25 \\ 1 & 1 & 1 & 1 & 0.25 & 0.25 & 0.25 & 0.25 \\ \sqrt{2} & -\sqrt{2} & \sqrt{2} & -\sqrt{2} & 0.354 & -0.354 & 0.354 & -0.354 \end{bmatrix}$$

Thus, we get

$$K(x_i, x_j) = \phi^T(x_i)\phi(x_j)$$

$$= \begin{bmatrix} 1 & 1 & \sqrt{2} \\ 1 & 1 & -\sqrt{2} \\ 1 & 1 & \sqrt{2} \\ 1 & 1 & -\sqrt{2} \\ 0.25 & 0.25 & 0.354 \\ 0.25 & 0.25 & -0.354 \\ 0.25 & 0.25 & 0.354 \\ 0.25 & 0.25 & -0.354 \end{bmatrix}$$

$$\times \begin{bmatrix} 1 & 1 & 1 & 1 & 0.25 & 0.25 & 0.25 & 0.25 \\ 1 & 1 & 1 & 1 & 0.25 & 0.25 & 0.25 & 0.25 \\ \sqrt{2} & -\sqrt{2} & \sqrt{2} & -\sqrt{2} & 0.354 & -0.354 & 0.354 & -0.354 \end{bmatrix}$$

$$= \begin{bmatrix} 4 & 0 & 4 & 0 & 1 & 0 & 1 & 0 \\ 0 & 4 & 0 & 4 & 0 & 1 & 0 & 1 \\ 4 & 0 & 4 & 0 & 1 & 0 & 1 & 0 \\ 0 & 4 & 0 & 4 & 0 & 1 & 0 & 1 \\ 1 & 0 & 1 & 0 & 0.25 & 0 & 0.25 & 0 \\ 0 & 1 & 0 & 1 & 0 & 0.25 & 0 & 0.25 \\ 1 & 0 & 1 & 0 & 0.25 & 0 & 0.25 & 0 \\ 0 & 1 & 0 & 1 & 0 & 0.25 & 0 & 0.25 \end{bmatrix}$$

The optimization problem in the dual form is given by
Maximize

$$Q(\alpha) = \sum_{i=1}^{N} \alpha_i - \frac{1}{2} \sum_{i=1}^{N} \sum_{j=1}^{N} \alpha_i \alpha_j y_i y_j K(x_i, x_j)$$

Subject to

$$\sum_{i=1}^{N} y_i \alpha_i = 0$$

$$\alpha_i \geq 0, i = 1, 2, ..., N.$$

By substituting the values of $K(x_i, x_j)$ and response variable y_i, the optimization problem in the dual form becomes

Maximize

$$\alpha_1 + \alpha_2 + \alpha_3 + \alpha_4 + \alpha_5 + \alpha_6 + \alpha_7 + \alpha_8 - \frac{1}{2}\left(4\alpha_1^2 + 4\alpha_2^2 + 4\alpha_3^2 + 4\alpha_4^2\right.$$

$$+ \frac{1}{4}\alpha_5^2 + \frac{1}{4}\alpha_6^2 + \frac{1}{4}\alpha_7^2 + \frac{1}{4}\alpha_8^2 + 8\alpha_1\alpha_3 - 2\alpha_1\alpha_5 - 2\alpha_1\alpha_7 + 8\alpha_2\alpha_4 - 2\alpha_2\alpha_6$$

$$\left. -2\alpha_2\alpha_8 - 2\alpha_3\alpha_5 - 2\alpha_3\alpha_7 - 2\alpha_4\alpha_6 - 2\alpha_4\alpha_8 + \frac{1}{2}\alpha_5\alpha_7 + \frac{1}{2}\alpha_6\alpha_8\right)$$

Subject to

$$\alpha_1 + \alpha_2 + \alpha_3 + \alpha_4 - \alpha_5 - \alpha_6 - \alpha_7 - \alpha_8 = 0$$

$$\alpha_i \geq 0, \quad i = 1, 2, \ldots, N.$$

The Lagrangian of this problem is

$$L(\alpha, \lambda) = \alpha_1 + \alpha_2 + \alpha_3 + \alpha_4 + \alpha_5 + \alpha_6 + \alpha_7 + \alpha_8 - \frac{1}{2}\left(4\alpha_1^2 + 4\alpha_2^2 + 4\alpha_3^2 + 4\alpha_4^2\right.$$

$$+ \frac{1}{4}\alpha_5^2 + \frac{1}{4}\alpha_6^2 + \frac{1}{4}\alpha_7^2 + \frac{1}{4}\alpha_8^2 + 8\alpha_1\alpha_3 - 2\alpha_1\alpha_5 - 2\alpha_1\alpha_7 + 8\alpha_2\alpha_4$$

$$-2\alpha_2\alpha_6 - 2\alpha_2\alpha_8 - 2\alpha_3\alpha_5 - 2\alpha_3\alpha_7 - 2\alpha_4\alpha_6 - 2\alpha_4\alpha_8$$

$$\left. + \frac{1}{2}\alpha_5\alpha_7 + \frac{1}{2}\alpha_6\alpha_8\right) + \lambda(\alpha_1 + \alpha_2 + \alpha_3 + \alpha_4 - \alpha_5 - \alpha_6 - \alpha_7 - \alpha_8)$$

$$\frac{\partial L}{\partial \alpha_1} = \frac{\partial L}{\partial \alpha_2} = \cdots = \frac{\partial L}{\partial \alpha_8} = \frac{\partial L}{\partial \lambda} = 0 \text{ gives the following equations,}$$

$$1 - 4\alpha_1 - 4\alpha_3 + \alpha_5 + \alpha_7 + \lambda = 0$$

$$1 - 4\alpha_2 - 4\alpha_4 + \alpha_6 + \alpha_8 + \lambda = 0$$

$$1 + \alpha_1 + \alpha_3 - \frac{1}{4}\alpha_5 - \frac{1}{4}\alpha_7 - \lambda = 0$$

$$1 + \alpha_2 + \alpha_4 - \frac{1}{4}\alpha_6 - \frac{1}{4}\alpha_8 - \lambda = 0$$

And

$$\alpha_1 + \alpha_2 + \alpha_3 + \alpha_4 - \alpha_5 - \alpha_6 - \alpha_7 - \alpha_8 = 0$$

The remaining four equations are redundant. So, a unique solution does not exist.

Suppose $\alpha_1 + \alpha_2 + \alpha_3 + \alpha_4 = u$ and $\alpha_5 + \alpha_6 + \alpha_7 + \alpha_8 = v$

With this notation, one can show that the above equations are reduced to

$$12u - 3v - 16 = 0$$

$$u = v$$

Solving we get,

$$u = v = \frac{16}{9}$$

Or, $\alpha_1 + \alpha_2 + \alpha_3 + \alpha_4 = \alpha_5 + \alpha_6 + \alpha_7 + \alpha_8 = \frac{16}{9}$.

Since the solution is not unique, to avoid mathematical complexity we can assume that, $\alpha_1 = \alpha_2 = \alpha_3 = \cdots = \alpha_8$ and the corresponding optimal values are $\alpha_1 = \alpha_2 = \alpha_3 = \cdots = \alpha_8 = \frac{4}{9}$.

The optimum values of the Lagrange multipliers are given by

$$\alpha_1^* = \alpha_2^* = \alpha_3^* = \alpha_4^* = \alpha_5^* = \alpha_6^* = \alpha_7^* = \alpha_8^* = \frac{4}{9}$$

Hence, the optimum value of the objective function $Q(\alpha)$ of the dual problem becomes

$$Q(\alpha^*) = \frac{4}{9} + \frac{4}{9} + \frac{4}{9} + \frac{4}{9} + \frac{4}{9} + \frac{4}{9} + \frac{4}{9} + \frac{4}{9} - \frac{1}{2} \times \left(4 \times \frac{4^2}{9} + 4 \times \frac{4^2}{9} \right.$$

$$+ 4 \times \frac{4^2}{9} + 4 \times \frac{4^2}{9} + \frac{1}{4} \times \frac{4^2}{9} + \frac{1}{4} \times \frac{4^2}{9} + \frac{1}{4} \times \frac{4^2}{9} + \frac{1}{4}$$

$$\times \frac{4^2}{9} + 8 \times \frac{4}{9} \times \frac{4}{9} - 2 \times \frac{4}{9} \times \frac{4}{9} - 2 \times \frac{4}{9} \times \frac{4}{9} + 8 \times \frac{4}{9} \times \frac{4}{9} - 2$$

$$\times \frac{4}{9} \times \frac{4}{9} - 2 \times \frac{4}{9} \times \frac{4}{9} - 2 \times \frac{4}{9} \times \frac{4}{9} - 2 \times \frac{4}{9} \times \frac{4}{9} - 2 \times \frac{4}{9} \times \frac{4}{9}$$

$$\left. - 2 \times \frac{4}{9} \times \frac{4}{9} + \frac{1}{2} \times \frac{4}{9} \times \frac{4}{9} + \frac{1}{2} \times \frac{4}{9} \times \frac{4}{9} \right) = 1.777$$

Also, we get

$$\sum_{i=1}^{N} \alpha_i^* = 8 \times \frac{4}{9}$$

$$= 3.554$$

In Eq. (3.20), by replacing x_i by $\phi(x_i)$, we can write

$$w^* = \sum_{i=1}^{N} y_i \alpha_i^* \phi(x_i)$$

Hence,

$$w^{*T}w^* = \sum_{i=1}^{N}\sum_{j=1}^{N}\alpha_i^*\alpha_j^*y_iy_j\phi^T(x_i)\phi(x_j)$$

$$= \sum_{i=1}^{N}\sum_{j=1}^{N}\alpha_i^*\alpha_j^*y_iy_jK(x_i,x_j)$$

Therefore, we can write

$$Q(\alpha^*) = \sum_{i=1}^{N}\alpha_i^* - \frac{1}{2}\sum_{i=1}^{N}\sum_{j=1}^{N}\alpha_i^*\alpha_j^*y_iy_jK(x_i,x_j)$$

$$= \sum_{i=1}^{N}\alpha_i^* - \frac{1}{2}w^{*T}w^*$$

$$\text{Or, } \frac{1}{2}w^{*T}w^* = \sum_{i=1}^{N}\alpha_i^* - Q(\alpha^*)$$

$$\text{Or, } \frac{1}{2}w^{*T}w^* = 3.554 - 1.777 = 1.777$$

$$\text{Or, } w^{*T}w^* = 3.554$$

Thus,

$$\|w^*\| = \sqrt{3.554} = 1.8852$$

The optimum weight vector is

$$w^* = \sum_{i=1}^{N}y_i\alpha_i^*\phi(x_i)$$

$$= \frac{4}{9}\{\phi(x_1) + \phi(x_2) + \phi(x_3) + \phi(x_4) - \phi(x_5) - \phi(x_6) - \phi(x_7) - \phi(x_8)\}$$

$$= \frac{4}{9}\left\{\begin{bmatrix}1\\1\\\sqrt{2}\end{bmatrix} + \begin{bmatrix}1\\1\\-\sqrt{2}\end{bmatrix} + \begin{bmatrix}1\\1\\\sqrt{2}\end{bmatrix} + \begin{bmatrix}1\\1\\-\sqrt{2}\end{bmatrix} - \begin{bmatrix}0.25\\0.25\\0.354\end{bmatrix} - \begin{bmatrix}0.25\\0.25\\-0.354\end{bmatrix}\right.$$

$$\left. - \begin{bmatrix}0.25\\0.25\\0.354\end{bmatrix} - \begin{bmatrix}0.25\\0.25\\-0.354\end{bmatrix}\right\} = \begin{bmatrix}1.33\\1.33\\0\end{bmatrix}$$

Thus, we have

$$w^{*T}\phi(x_i) = [1.33 \ \ 1.33 \ \ 0]$$

$$\times \begin{bmatrix} 1 & 1 & 1 & 1 & 0.25 & 0.25 & 0.25 & 0.25 \\ 1 & 1 & 1 & 1 & 0.25 & 0.25 & 0.25 & 0.25 \\ \sqrt{2} & -\sqrt{2} & \sqrt{2} & -\sqrt{2} & 0.354 & -0.354 & 0.354 & -0.354 \end{bmatrix}$$

$$= [2.67 \ \ 2.67 \ \ 2.67 \ \ 2.67 \ \ 0.67 \ \ 0.67 \ \ 0.67 \ \ 0.67]$$

Obviously,

$$\min\left(w^{*T}\phi(x_i)\right) = 2.67, \quad \text{when } y_i = +1$$

and

$$\max\left(w^{*T}\phi(x_i)\right) = 0.67, \quad \text{when } y_i = -1$$

Hence, the optimum value of b is

$$b^* = -\frac{\displaystyle\max_{y_i=-1}\left(w^{*T}\phi(x_i)\right) + \min_{y_i=+1}\left(w^{*T}\phi(x_i)\right)}{2}$$

$$= -\frac{0.67 + 2.67}{2}$$

$$= -1.67$$

Therefore, the equation of optimal hyperplane is

$$w^{*T}\phi(x_i) + b^* = 0$$

$$\text{Or, } [1.33 \ \ 1.330] \begin{bmatrix} x_1^2 \\ x_2^2 \\ \sqrt{2}x_1x_2 \end{bmatrix} - 1.667 = 0$$

$$\text{Or, } 1.33x_1^2 + 1.33x_2^2 - 1.67 = 0$$

$$\text{Or, } x_1^2 + x_2^2 = 1.12^2 \tag{3.33}$$

Also, the equation of the first margin is

$$w^{*T}\phi(x_i) + b^* = 1$$

$$\text{Or, } 1.33x_1^2 + 1.33x_2^2 - 1.67 = 1$$

$$\text{Or, } x_1^2 + x_2^2 = 1.42^2 \tag{3.34}$$

Similarly, the equation of the second margin is

$$w^{*T}\phi(x_i) + b^* = -1$$

$$\text{Or, } 1.33x_1^2 + 1.33x_2^2 - 1.67 = -1$$

$$\text{Or, } x_1^2 + x_2^2 = 0.71^2 \tag{3.35}$$

Eqs. (3.33)–(3.35) are the equations of circles with radiuses 1.12, 1.42, and 0.71, respectively. Fig. 3.7 depicts the optimum separating hyperplane with two margins. Here, all eight data points are support vectors as all of them are lying on the margins. Fig. 3.8 shows that the data points are linearly separable in high-dimensional feature space.

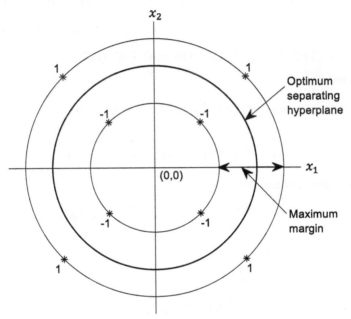

FIG. 3.7 Nonlinear classification in two dimensions.

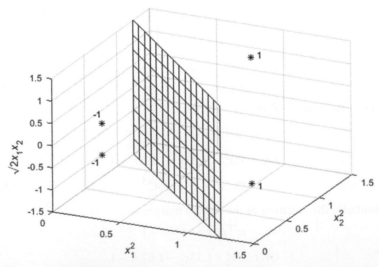

FIG. 3.8 Linear classification in three-dimensional feature space.

3.2.3 Application of SVC for binary classification: Identification of handloom and power loom fabrics

Handloom fabrics differ from the power loom ones in terms of uniformity of pick spacing, crimp evenness, cover variation, etc. (Ghosh et al., 2015). Handloom fabrics being woven manually often encounter variations in the force applied during the insertion of picks. Further, uneven pressure on treadle pedals during manual weaving may contribute to the formation of sheds where warp threads are inconsistently tensioned. All such variations translate themselves in the formation of a fabric characterized by an uneven rugged appearance as opposed to a power loom woven product which is far more even and uniform in appearance in spite of the fact that they are woven from the same yarns. This very appearance confers upon it an attribute that earns a rare ethnic appeal in its otherwise rugged texture that a similar power loom product with identical structural and raw material specifications can hardly match. In reality, the handloom products are highly adored all over the world. Therefore, justifiably it does carry a premium price tag. Cheap inferior imitations from the power loom sector are eating into its pie of profits precipitating an ailing handloom sector. Need of the hour is therefore an effective automatic recognition mechanism making a distinction between handloom fabrics and their power loom counterparts. With this urge before us, we present here an approach using SVC to realize this distinction.

A pattern recognition system for classifying handloom and power loom woven fabrics can be partitioned into a number of components as illustrated in Fig. 3.9. At first, a digital camera captures the images of handloom and power loom fabrics. Next, the camera's signals are processed to simplify subsequent operations without losing relevant information. The information from each

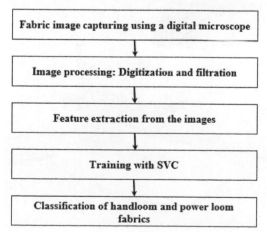

FIG. 3.9 Flowchart of the pattern recognition system for classifying handloom and power loom fabrics.

fabric image is then sent to a feature extractor, whose purpose is to reduce the data by measuring certain features or attributes. SVC uses these features to evaluate the evidence presented and make a final decision as to the fabric type.

Plain woven fabrics were prepared from 20 Ne warp and 30 Ne weft both in handloom and power loom. Fabric images were captured using a LEICA camera (Model EZ-4D) with a magnification of 25×. Samples were illuminated by three halogen lights positioned approximately 20 cm above directly and to the right and left of the sample to supply illumination in diagonal directions of 45°. The digitized images were constituted of 1024×768 pixels with subsequent conversion into gray level of 0–255 and stored as a two-dimensional gray matrix. Fig. 3.10 depicts the digital microscope used for image capturing. All images were taken under identical conditions of magnification and illumination.

For feature extraction, digitized gray images of handloom and power loom fabrics were considered. Fig. 3.11 depicts typical images of handloom and power loom fabrics. The values of the mean intensities as obtained from the data matrix of a gray fabric image for each column and each row correspond to the signals of warp and weft directions, respectively. A plot of these signals both in warp and weft directions of the fabric as depicted in Fig. 3.12 reveals the presence of periodic peaks the positions of which lie approximately on the axes of warp and weft threads in the image. Thus, for every image a grid was constructed out of a set of horizontal and vertical axes whose every point of intersection is the position of interlacement of warp and weft.

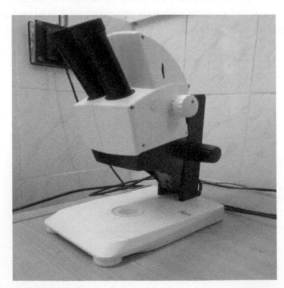

FIG. 3.10 Camera used for capturing images of fabric.

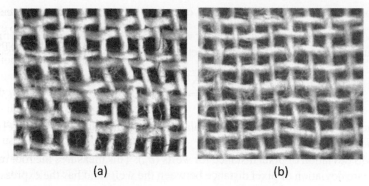

FIG. 3.11   Images of (A) handloom fabric and (B) power loom fabric.

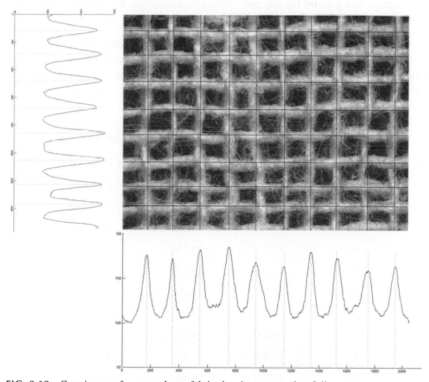

FIG. 3.12   Gray image of a power loom fabric showing warp and weft lines.

From each image, the following two feature parameters were extracted:

1. RMS deviation of pixel distance for warps (σ_e): This measures the root mean square deviation of pixel distance between the warps and has the expression:

$$\sigma_e = \sqrt{\frac{\sum (e_i - e)^2}{n - 1}} \qquad (3.36)$$

where e_i and e stand for any particular pixel distance and average pixel distance for warps, respectively.

2. RMS deviation of pixel distance for wefts (σ_p): This measures the root mean square deviation of pixel distance between the wefts and has the expression:

$$\sigma_p = \sqrt{\frac{\sum (p_i - p)^2}{n - 1}} \qquad (3.37)$$

where p_i and p stand for any particular pixel distance and average pixel distance for wefts respectively.

The dataset comprises a total of 50 data points encompassing 25 each for handloom and power loom fabrics. Table 3.4 refers to the datasets representing two different features for both the two classes. The first 25 datasets correspond to handloom fabrics and the rest of the dataset refers to power loom fabrics.

TABLE 3.4 Extracted features from the fabric images.

Sl. no.	σ_e	σ_p	Class
1	18.635	54.008	1
2	17.332	83.395	1
3	19.302	56.474	1
4	17.713	38.898	1
5	14.397	55.843	1
6	12.726	57.288	1
7	51.872	44.926	1
8	30.331	65.963	1
9	29.348	67.969	1
10	27.231	50.54	1
11	22.611	45.371	1
12	27.696	44.358	1
13	14.618	50.691	1

TABLE 3.4 Extracted features from the fabric images—cont'd

Sl. no.	σ_e	σ_p	Class
14	40.119	59.768	1
15	23.064	44.229	1
16	10.59	46.401	1
17	51.293	59.983	1
18	39.611	49.976	1
19	16.771	50.425	1
20	35.978	44.431	1
21	30.615	53.942	1
22	17.096	44.415	1
23	24.487	66.153	1
24	26.823	36.617	1
25	25.505	50.805	1
26	11.942	9.8894	−1
27	7.6811	7.8098	−1
28	11.487	9.0584	−1
29	13.479	5.7755	−1
30	11.303	14.093	−1
31	11.606	17.433	−1
32	14.937	13.507	−1
33	18.44	17.199	−1
34	9.6977	8.9249	−1
35	18.345	11.234	−1
36	7.3993	10.518	−1
37	21.398	18.8	−1
38	25.536	10.706	−1
39	23.929	13.37	−1
40	16.476	14.566	−1
41	14.821	15.221	−1
42	11.708	13.201	−1

Continued

TABLE 3.4 Extracted features from the fabric images—cont'd

Sl. no.	σ_e	σ_p	Class
43	11.023	16.858	−1
44	15.263	14.557	−1
45	8.5294	11.236	−1
46	10.712	11.41	−1
47	9.7852	13.458	−1
48	9.7468	16.161	−1
49	12.994	12.045	−1
50	12.587	17.842	−1

A soft margin SVC with a linear kernel function has been used for the classification of handloom and power loom fabric images. A sequence of arithmetic progress with a common difference of 5, such as 5th, 10th, 15th, …, 50th data points, is used as a testing dataset, and remaining 40 data points are used as a training dataset. The mean absolute percentage accuracy between the actual and predicted values for both training and testing sets is found to be 100%. Fig. 3.13

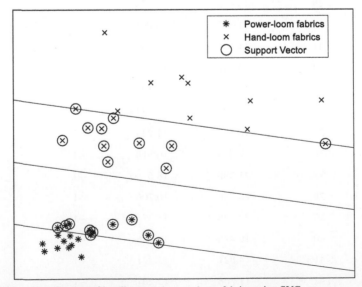

FIG. 3.13 Classification of handloom and power loom fabrics using SVC.

illustrates the optimum hyperplane and margins for classifying the handloom and power loom fabrics. Handloom fabrics are marked with × symbol and power loom fabrics are marked with * symbol.

The present study holds the key to an effective checking mechanism to differentiate handloom and power loom products to protect the interest of both customers and poor handloom weavers.

3.2.4 Application of SVC for multi-class classification: Identification of various fabric defects

Tsai et al. (1995) approached the fabric defect identification problem by applying conventional multilayer perception using the backpropagation algorithm. The cloth sample of their experimentation was polyester/wool blended twill weave with 92 and 42 ends/in and picks/in, respectively. They selected four classes of defects for identification viz. neps, broken end, broken pick, and oil stain. These classes were identified as class-1 for normal, class-2 for neps, class-3 for broken end, class-4 for broken pick, and class-5 for oil stain. Fig. 3.14 shows images of such defects.

Ghosh et al. (2011) used Tsai et al. (1995) experimental data for the purpose of classifying different fabric defects using the SVC with the Gaussian radial

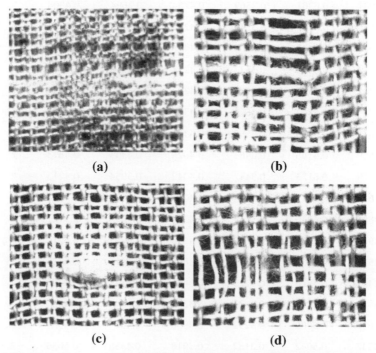

(a)   (b)

(c)   (d)

FIG. 3.14 Typical image of different fabric defects (A) oil stain, (B) broken end, (C) neps, and (D) broken pick.

basis kernel function. A gray-level co-occurrence matrix was employed to obtain the feature parameters f_1, f_2, f_3, f_4, f_5, f_6, for various classes. Among the feature vectors, f_1, f_2, f_3, and f_4 are the contrast measurement of texture images along the direction angle (θ) of 0°, 45°, 90°, and 135°, respectively with spatial displacement $d = 1$. After carefully analyzing the graphic representation depicting contrast values in relation to pixel distance it was found that the strongest response at pixel distance 12 along horizontal direction and 16 along vertical direction in case of normal fabrics. Thus, it was further included two more features f_5 and f_6 as contrast values at $d = 12$, $\theta = 0°$, $d = 16$, $\theta = 90°$, respectively. Table 3.5 shows the complete data set which comprises a total of 50 experimental data encompassing 10 experiments per class. The concept of

TABLE 3.5 Dataset for various kinds of fabric defects.

f_1	f_2	f_3	f_4	f_5	f_6	Class
0.39	0.6402	0.3584	0.4205	0.3726	0.3434	1
0.4026	0.6362	0.3601	0.432	0.3438	0.3442	1
0.3879	0.6161	0.3419	0.4153	0.3228	0.3547	1
0.3931	0.6381	0.3569	0.4284	0.3694	0.4308	1
0.3826	0.6298	0.3537	0.4234	0.3489	0.3435	1
0.3978	0.6433	0.3704	0.443	0.3584	0.3811	1
0.392	0.6464	0.3532	0.4221	0.3352	0.3859	1
0.3887	0.6363	0.3601	0.4202	0.322	0.3257	1
0.388	0.6322	0.3672	0.4302	0.3481	0.3378	1
0.3851	0.6228	0.3567	0.4361	0.3496	0.3371	1
0.3689	0.6188	0.3483	0.4026	0.4393	0.4813	2
0.3789	0.6173	0.3447	0.4042	0.3954	0.4213	2
0.3663	0.6173	0.3444	0.4045	0.4439	0.4788	2
0.3881	0.6345	0.3569	0.4305	0.4214	0.5121	2
0.3964	0.6362	0.3512	0.4236	0.4049	0.421	2
0.3529	0.5768	0.3219	0.3865	0.4417	0.4725	2
0.3465	0.5874	0.3225	0.3819	0.474	0.5255	2
0.3467	0.5767	0.313	0.3782	0.3845	0.4925	2
0.3697	0.5805	0.3232	0.3978	0.466	0.4953	2
0.3537	0.5642	0.3182	0.3918	0.4358	0.5035	2

TABLE 3.5 Dataset for various kinds of fabric defects—cont'd

f_1	f_2	f_3	f_4	f_5	f_6	Class
0.3509	0.5957	0.3507	0.4079	0.5432	0.3107	3
0.3661	0.5915	0.3361	0.4137	0.4808	0.2884	3
0.3717	0.5968	0.3237	0.4003	0.4708	0.3376	3
0.3589	0.5903	0.323	0.3931	0.4377	0.3266	3
0.3436	0.5775	0.3298	0.3907	0.4888	0.3454	3
0.3159	0.5158	0.3214	0.3981	0.5433	0.3301	3
0.3354	0.5356	0.3373	0.4095	0.5594	0.3677	3
0.3231	0.5202	0.3197	0.3899	0.5466	0.351	3
0.3534	0.5655	0.3275	0.4129	0.521	0.3302	3
0.3761	0.5795	0.3399	0.4324	0.529	0.3305	3
0.3723	0.5821	0.2097	0.3695	0.3453	0.3765	4
0.3836	0.6022	0.3054	0.3861	0.3383	0.3429	4
0.3716	0.5918	0.3101	0.3761	0.3595	0.3248	4
0.4115	0.6037	0.2797	0.4036	0.3987	0.3294	4
0.4321	0.6446	0.309	0.4157	0.4254	0.3284	4
0.3765	0.608	0.3098	0.3842	0.3198	0.3587	4
0.3987	0.6132	0.3145	0.3954	0.3272	0.3829	4
0.384	0.5953	0.3123	0.392	0.3165	0.4022	4
0.3854	0.6023	0.3101	0.389	0.3154	0.3635	4
0.3873	0.597	0.3074	0.3944	0.3554	0.3735	4
0.4	0.4976	0.3254	0.3969	0.5242	0.4233	5
0.2626	0.3115	0.2417	0.2633	0.4584	0.3841	5
0.2657	0.3276	0.2263	0.2723	0.3681	0.4321	5
0.364	0.4823	0.3034	0.3518	0.5274	0.62	5
0.4051	0.5158	0.3361	0.4082	0.6228	0.6095	5
0.3592	0.4453	0.3003	0.3543	0.4673	0.41	5
0.4049	0.4874	0.3207	0.3977	0.5187	0.424	5
0.3586	0.4805	0.3102	0.3614	0.4967	0.8066	5
0.3049	0.3866	0.2726	0.3215	0.4967	0.5492	5
0.4029	0.5257	0.3363	0.4028	0.5465	0.4661	5

binary classification using SVC was extended into multi-category classes in context by ensuing the principle of one versus the rest and constructing 10 binary SVC classifiers with the following pair-wise classification:

class-1 vs class-2
class-1 vs class-3
class-1 vs class-4
class-1 vs class-5
class-2 vs class-3
class-2 vs class-4
class-2 vs class-5
class-3 vs class-4
class-3 vs class-5
class-4 vs class-5

This agrees with the principle of pair-wise classification that returns the number of binary classifiers as 10, resulting from the formula $N(N-1)/2$ where N is the total number of class labels.

Among quite a few techniques of cross-validation (Han & Kamber, 2006), "one-leave-out" method has been used here to evaluate the performance of the SVC. In this method of cross-validation, one sample is left out at a time for the testing and the rest of the dataset are used for the training. And for the next cycle a new sample is taken for the testing and rest of the dataset are used as training dataset. The cycles repeat until each sample is used the same number of times for training and once for testing. Hence the total number of cycles is equal to the size of the dataset. The one-leave-out cross-validation method for n number of data sets is schematically depicted in Fig. 3.15. We thereafter evaluate training and testing accuracy in every cycle and estimate the overall training and testing accuracies which are defined as follows:

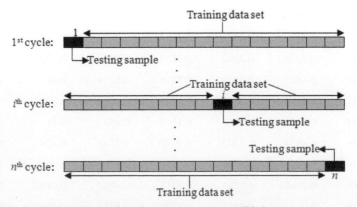

FIG. 3.15 Schematic representation of one-leave-out-cross-validation.

$$\text{Training accuracy } \% = \frac{\text{Number of accurately predicted training data points}}{\text{Total number of training data points}} \times 100$$

$$\text{Testing accuracy } \% = \frac{\text{Number of accurately predicted testing data points}}{\text{Total number of testing data points}} \times 100$$

As the dataset contains a total of 50 data points, the SVC was trained using 49 data and tested on the sample left out for each cycle and this continued for 50 cycles. The overall training and testing accuracies were both found to be 100%. Robust algorithm of SVC confers upon it the ability to treat a great volume of data having multi-dimensions with unwavering performance and this holds hope to extend such classifier to recognition task of many more classes of fabric defects as a real-time online automatic fabric defect inspection system.

3.3 Support vector regression (SVR)

In a regression problem, an algorithm is constructed that estimates an unknown mapping between a system's input and output parameters from the available dataset. On estimating such an input-output relation, the model can be used to predict system outputs from the given inputs for a new set of data points. Consider a training data set $S = \{[x_i, y_i] \in \mathfrak{R}^n \times \mathfrak{R}, i = 1, 2, ..., l\}$ consists of l pairs $(x_1, y_1), (x_2, y_i), ..., (x_l, y_l)$, where the inputs $x \in \mathfrak{R}^n$ are n-dimensional vectors and the system's responses $y \in \mathfrak{R}$ are continuous values, $\mathfrak{R}$ and $\mathfrak{R}^n$ denote the real field and Euclidean space of the real number system of n-dimension, respectively. In support vector regression (SVR), the aim is to obtain a function from the training data set S in the following form

$$f(x) = w^T x_i + b \tag{3.38}$$

which approximates best the system response, where w denotes the weight vector and b is the bias term. The parameters w and b in Eq. (3.38) are estimated using the training set S. In order to ensuring the existence of a global minimum as well as improving the generalization accuracy, Vapnik (1995) introduced the concept of ε-insensitive loss function, which has the following form

$$L^\varepsilon(x, y, f) = |y - f(x)|_\varepsilon = \max(0, |y - f(x)| - \varepsilon) \tag{3.39}$$

where f is a real-valued function on the input domain X, $x \in X$ and $y \in \mathfrak{R}$ and ε is a precision parameter representing the radius of the tube located around the regression function. The SVR algorithm attempts to position the ε tube around the function $f(x)$. Fig. 3.16 depicts an example of a one-dimensional linear regression function with an ε-insensitive band. The slack variables ξ and ξ' in Fig. 3.16 measures the cost of the errors on the training points. These are zero for all points inside the band. Fig. 3.17 demonstrates the form of linear

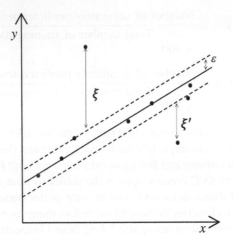

FIG. 3.16   The insensitive band for a one-dimensional linear regression problem.

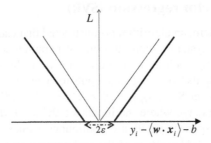

FIG. 3.17   The linear ε-insensitive loss for zero and non-zero ε.

ε-insensitive losses for zero and non-zero ε as a function of $y - f(x)$. Vapnik (1995) showed that for optimizing the generalization performance of the linear regressor, we should

$$\text{Minimize } \frac{1}{2}\|w\|^2 + C\sum_{i=1}^{l}\left(\xi_i + \xi_i'\right) \tag{3.40}$$

Subject to

$$y_i - \left(w^T x_i + b\right) \le \varepsilon + \xi_i$$

$$\left(w^T x_i + b\right) - y_i \le \varepsilon + \xi_i'$$

$$\xi_i \ge 0, i = 1, 2, \dots l.$$

$$\xi_i' \ge 0, i = 1, 2, \dots, l.$$

where C is a prespecified value, and ξ_i, ξ_i' are slack variables representing upper and lower constraints on the outputs of the system. The corresponding Lagrangian function for this optimization problem is given by

$$L(w,b,\xi,\xi',\alpha,\alpha',\beta,\beta') = \frac{1}{2}\|w\|^2 + C\sum_{i=1}^{l}(\xi_i + \xi_i')$$

$$- \sum_{i=1}^{l}\alpha_i\left[w^T x_i + b - y_i + \varepsilon + \xi_i\right]$$

$$- \sum_{i=1}^{l}\alpha_i'\left[y_i - \left(w^T x_i + b\right) + \varepsilon + \xi_i'\right] \tag{3.41}$$

$$- \sum_{i=1}^{l}\beta_i\xi_i - \sum_{i=1}^{l}\beta_i'\xi_i'$$

with $\alpha_i \geq 0$, $\alpha_i' \geq 0$, $\beta_i \geq 0$, and $\beta_i' \geq 0$. The coefficients α_i, α_i', β_i, and β_i' are called the Lagrange multipliers. The corresponding dual is found by differentiating with respect to w, b, ξ, ξ', and imposing stationarity we have

$$\frac{\partial L(w,b,\xi,\xi',\alpha,\alpha',\beta,\beta')}{\partial w} = w - \sum_{i=1}^{l}\alpha_i x_i + \sum_{i=1}^{l}\alpha_i' x_i = 0 \tag{3.42}$$

$$\frac{\partial L(w,b,\xi,\xi',\alpha,\alpha',\beta,\beta')}{\partial b} = -\sum_{i=1}^{l}\alpha_i + \sum_{i=1}^{l}\alpha_i' = 0 \tag{3.43}$$

$$\frac{\partial L(w,b,\xi,\xi',\alpha,\alpha',\beta,\beta')}{\partial \xi} = C - \sum_{i=1}^{l}\alpha_i - \sum_{i=1}^{l}\beta_i = 0 \tag{3.44}$$

$$\frac{\partial L(w,b,\xi,\xi',\alpha,\alpha',\beta,\beta')}{\partial \xi'} = C - \sum_{i=1}^{l}\alpha_i' - \sum_{i=1}^{l}\beta_i' = 0 \tag{3.45}$$

and resubstituting the relations obtained into the primal with some simplification we arrive at the following adaptation of the dual objective function:

$$L(w,b,\xi,\xi',\alpha,\alpha',\beta,\beta') = \sum_{i=1}^{l}(\alpha_i - \alpha_i')y_i - \varepsilon\sum_{i=1}^{l}(\alpha_i + \alpha_i')$$

$$- \frac{1}{2}\sum_{i,j=1}^{l}(\alpha_i - \alpha_i')\left(\alpha_j - \alpha_j'\right)x_i^T x_j \tag{3.46}$$

Hence, maximizing the above objective function over α_i and α_i' is equivalent to

Maximize

$$Q(\alpha) = \sum_{i=1}^{l}(\alpha_i - \alpha_i')y_i - \varepsilon\sum_{i=1}^{l}(\alpha_i + \alpha_i')$$

$$- \frac{1}{2}\sum_{i,j=1}^{l}(\alpha_i - \alpha_i')\left(\alpha_j - \alpha_j'\right)x_i^T x_j \tag{3.47}$$

Subject to

$$0 \leq \alpha_i, \quad \alpha_i' \leq C$$

$$\sum_{i=1}^{l}(\alpha_i - \alpha_i') = 0, \quad i = 1, 2, ..., l.$$

The corresponding Karush-Kuhn-Tucker (KKT) complementarity conditions are

$$\alpha_i\left(w^T x_i + b - y_i + \varepsilon + \xi_i\right) = 0$$
$$\alpha_i'\left(y_i - w^T x_i - b + \varepsilon + \xi_i'\right) = 0$$
$$\xi_i \xi_i' = 0$$
$$\alpha_i \alpha_i' = 0$$
$$(\alpha_i - C)\xi_i = 0$$
$$(\alpha_i' - C)\xi_i' = 0, \quad i = 1, 2 \ldots, l.$$

Solving Eq. (3.47) with constraints determines the Lagrange multipliers, α_i, α_i'. From Eq. (3.42), the optimum weight vector becomes

$$w^* = \sum_{i=1}^{l}\left(\alpha_i^* - \alpha_i'^*\right)x_i \tag{3.48}$$

Thus, the optimum regression function is given by

$$f(x) = w^{*T} x_i + b^*$$
$$= \sum_{i=1}^{l}\left(\alpha_i^* - \alpha_i'^*\right)x_i^T x_j + b^* \tag{3.49}$$

where b^* is expressed as

$$b^* = \begin{cases} y_i - w^{*T} x - \epsilon, & \text{for } 0 < \alpha_i^* < C \\ y_i - w^{*T} x + \epsilon, & \text{for } 0 < \alpha_i'^* < C \end{cases} \tag{3.50}$$

Using the Kernel function, Eq. (3.48) is modified as

$$f(x) = \sum_{i=1}^{l}\left(\alpha_i^* - \alpha_i'^*\right)K\left(x_i, x_j\right) + b^* \tag{3.51}$$

Using the relation $\alpha_i \alpha_i' = 0$ of the KKT conditions, the support vectors (sv) are points where exactly one of the Lagrange multipliers is greater than zero. Therefore, if we consider the band of $\pm\varepsilon$ around the function output by the learning algorithm, the points that are not strictly inside the tube are sv and those not touching the tube will have the absolute value of that parameter equal to C. Fig. 3.18 illustrates an example of a feature mapping, where linear regression of data cannot be possible in the input space but can be achieved in the feature space.

3.3.1 Step-by-step working principle of SVR

To illustrate the step-by-step procedure of a linear support vector regression model, we consider the training dataset as given in Table 3.6.

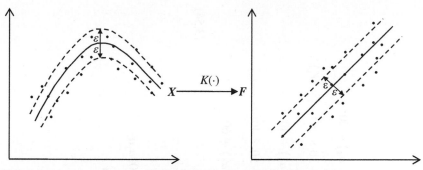

FIG. 3.18 A feature map from input space to higher dimensional feature space.

TABLE 3.6 Regression training set.

x	y
20	2.5
30	3.6
40	5
50	7
60	7.5

The optimization problem in the dual form is given by
Maximize

$$Q(\alpha) = \sum_{i=1}^{l}(\alpha_i - \alpha_i')y_i - \epsilon \sum_{i=1}^{l}(\alpha_i + \alpha_i') - \frac{1}{2}\sum_{i=1}^{l}\sum_{j=1}^{l}(\alpha_i - \alpha_i')(\alpha_j - \alpha_j')x_i^T x_j$$

Subject to

$$\sum_{i=1}^{N}(\alpha_i - \alpha_i') = 0$$

$$0 \le \alpha_i, \alpha_i' \le C, \quad i = 1, 2, ..., l.$$

We assume $\epsilon = 0.1$.
Now, from the given training dataset we have

$$\sum_{i=1}^{l}(\alpha_i - \alpha_i')y_i = 2.5(\alpha_1 - \alpha_1') + 3.6(\alpha_2 - \alpha_2') + 5(\alpha_3 - \alpha_3') + 7(\alpha_4 - \alpha_4')$$
$$+ 7.5(\alpha_5 - \alpha_5')$$

(3.52)

$$-\epsilon \sum_{i=1}^{l}(\alpha_i + \alpha_i') = -0.1(\alpha_1 + \alpha_1' + \alpha_2 + \alpha_2' + \alpha_3 + \alpha_3' + \alpha_4 + \alpha_4' + \alpha_5 + \alpha_5')$$

(3.53)

and

$$-\frac{1}{2}\sum_{i=1}^{l}\sum_{j=1}^{l}(\alpha_i - \alpha'_i)(\alpha_j - \alpha'_j)x_i x_j$$

$$= -\frac{1}{2}\Big[(\alpha_1 - \alpha'_1)\{20\times20(\alpha_1 - \alpha'_1) + 20\times30(\alpha_2 - \alpha'_2) + 20\times40(\alpha_3 - \alpha'_3) + 20\times50(\alpha_4 - \alpha'_4) + 20$$

$$\times60(\alpha_5 - \alpha'_5)\} + (\alpha_2 - \alpha'_2)\{30\times20(\alpha_1 - \alpha'_1) + 30\times30(\alpha_2 - \alpha'_2) + 30\times40(\alpha_3 - \alpha'_3) + 30\times50(\alpha_4 - \alpha'_4)$$

$$+30\times60(\alpha_5 - \alpha'_5)\} + (\alpha_3 - \alpha'_3)\{40\times20(\alpha_1 - \alpha'_1) + 40\times30(\alpha_2 - \alpha'_2) + 40\times40(\alpha_3 - \alpha'_3) + 40$$

$$\times50(\alpha_4 - \alpha'_4) + 40\times60(\alpha_5 - \alpha'_5)\} + (\alpha_4 - \alpha'_4)\{50\times20(\alpha_1 - \alpha'_1) + 50\times30(\alpha_2 - \alpha'_2) + 50\times40(\alpha_3 - \alpha'_3)$$

$$+50\times50(\alpha_4 - \alpha'_4) + 50\times60(\alpha_5 - \alpha'_5)\} + (\alpha_5 - \alpha'_5)\{60\times20(\alpha_1 - \alpha'_1) + 60\times30(\alpha_2 - \alpha'_2) + 60$$

$$\times40(\alpha_3 - \alpha'_3) + 60\times50(\alpha_4 - \alpha'_4) + 60\times60(\alpha_5 - \alpha'_5)\}\Big] \tag{3.54}$$

$$= -\frac{1}{2}\Big[400\alpha_1^2 + 400\alpha_1'^2 + 900\alpha_2^2 + 900\alpha_2'^2 + 1600\alpha_3^2 + 1600\alpha_3'^2 + 2500\alpha_4^2 + 2500\alpha_4'^2 + 3600\alpha_5^2$$

$$+ 3600\alpha_5'^2 + 2\Big\{-400\alpha_1\alpha'_1 - 900\alpha_2\alpha'_2 - 1600\alpha_3\alpha'_3 - 2500\alpha_4\alpha'_4 - 3600\alpha_5\alpha'_5 + 600\alpha_1\alpha_2 - 600\alpha'_1\alpha_2$$

$$-600\alpha_1\alpha'_2 + 600\alpha'_1\alpha'_2 + 800\alpha_1\alpha_3 - 800\alpha'_1\alpha_3 - 800\alpha_1\alpha'_3 + 800\alpha'_1\alpha'_3 + 1000\alpha_1\alpha_4 - 1000\alpha'_1\alpha_4$$

$$-1000\alpha_1\alpha'_4 + 1000\alpha'_1\alpha'_4 + 1200\alpha_1\alpha_5 - 1200\alpha'_1\alpha_5 - 1200\alpha_1\alpha'_5 + 1200\alpha'_1\alpha'_5 + 1200\alpha_2\alpha_3$$

$$-1200\alpha'_2\alpha_3 - 1200\alpha_2\alpha'_3 + 1200\alpha'_2\alpha'_3 + 1500\alpha_2\alpha_4 - 1500\alpha'_2\alpha_4 - 1500\alpha_2\alpha'_4 + 1500\alpha'_2\alpha'_4$$

$$+1800\alpha_2\alpha_5 - 1800\alpha'_2\alpha_5 - 1800\alpha_2\alpha'_5 + 1800\alpha'_2\alpha'_5 + 2000\alpha_3\alpha_4 - 2000\alpha'_3\alpha_4 - 2000\alpha_3\alpha'_4$$

$$+2000\alpha'_3\alpha'_4 + 2400\alpha_3\alpha_5 - 2400\alpha'_3\alpha_5 - 2400\alpha_3\alpha'_5 + 2400\alpha'_3\alpha'_5 + 3000\alpha_4\alpha_5 - 3000\alpha'_4\alpha_5$$

$$-3000\alpha_4\alpha'_5 + 3000\alpha'_4\alpha'_5\Big\}\Big]$$

By (3.52)+(3.53)+(3.54), we have

$$Q(\alpha) = 2.4\alpha_1 - 2.6\alpha_1' + 3.5\alpha_2 - 3.7\alpha_2' + 4.9\alpha_3 - 5.1\alpha_3' + 6.9\alpha_4 - 7.1\alpha_4'$$
$$+ 7.4\alpha_5 - 7.6\alpha_5' - \frac{1}{2}\Big[400\alpha_1^2 + 400\alpha_1'^2 + 900\alpha_2^2 + 900\alpha_2'^2 + 1600\alpha_3^2$$
$$+ 1600\alpha_3'^2 + 2500\alpha_4^2 + 2500\alpha_4'^2 + 3600\alpha_5^2 + 3600\alpha_5'^2$$
$$+ 2\Big\{-400\alpha_1\alpha_1' - 900\alpha_2\alpha_2' - 1600\alpha_3\alpha_3' - 2500\alpha_4\alpha_4' - 3600\alpha_5\alpha_5'$$
$$+ 600\alpha_1\alpha_2 - 600\alpha_1'\alpha_2 - 600\alpha_1\alpha_2' + 600\alpha_1'\alpha_2' + 800\alpha_1\alpha_3$$
$$- 800\alpha_1'\alpha_3 - 800\alpha_1\alpha_3' + 800\alpha_1'\alpha_3' + 1000\alpha_1\alpha_4 - 1000\alpha_1'\alpha_4$$
$$- 1000\alpha_1\alpha_4' + 1000\alpha_1'\alpha_4' + 1200\alpha_1\alpha_5 - 1200\alpha_1'\alpha_5 - 1200\alpha_1\alpha_5'$$
$$+ 1200\alpha_1'\alpha_5' + 1200\alpha_2\alpha_3 - 1200\alpha_2'\alpha_3 - 1200\alpha_2\alpha_3' + 1200\alpha_2'\alpha_3'$$
$$+ 1500\alpha_2\alpha_4 - 1500\alpha_2'\alpha_1 - 1500\alpha_2\alpha_4' + 1500\alpha_2'\alpha_4' + 1800\alpha_2\alpha_5$$
$$- 1800\alpha_2'\alpha_5 - 1800\alpha_2\alpha_5' + 1800\alpha_2'\alpha_5' + 2000\alpha_3\alpha_4 - 2000\alpha_3'\alpha_4$$
$$- 2000\alpha_3\alpha_4' + 2000\alpha_3'\alpha_4' + 2400\alpha_3\alpha_5 - 2400\alpha_3'\alpha_5 - 2400\alpha_3\alpha_5'$$
$$+ 2400\alpha_3'\alpha_5' + 3000\alpha_4\alpha_5 - 3000\alpha_4'\alpha_5 - 3000\alpha_4\alpha_5' + 3000\alpha_4'\alpha_5'\Big\}\Big]$$

Thus, the optimization problem in the dual form can be expressed as follows:
Maximize

$$Q(\alpha) = 2.4\alpha_1 - 2.6\alpha_1' + 3.5\alpha_2 - 3.7\alpha_2' + 4.9\alpha_3 - 5.1\alpha_3' + 6.9\alpha_4 - 7.1\alpha_4'$$
$$+ 7.4\alpha_5 - 7.6\alpha_5' - \frac{1}{2}\Big[400\alpha_1^2 + 400\alpha_1'^2 + 900\alpha_2^2 + 900\alpha_2'^2 + 1600\alpha_3^2$$
$$+ 1600\alpha_3'^2 + 2500\alpha_4^2 + 2500\alpha_4'^2 + 3600\alpha_5^2 + 3600\alpha_5'^2$$
$$+ 2\Big\{-400\alpha_1\alpha_1' - 900\alpha_2\alpha_2' - 1600\alpha_3\alpha_3' - 2500\alpha_4\alpha_4' - 3600\alpha_5\alpha_5'$$
$$+ 600\alpha_1\alpha_2 - 600\alpha_1'\alpha_2 - 600\alpha_1\alpha_2' + 600\alpha_1'\alpha_2' + 800\alpha_1\alpha_3$$
$$- 800\alpha_1'\alpha_3 - 800\alpha_1\alpha_3' + 800\alpha_1'\alpha_3' + 1000\alpha_1\alpha_4 - 1000\alpha_1'\alpha_4$$
$$- 1000\alpha_1\alpha_4' + 1000\alpha_1'\alpha_4' + 1200\alpha_1\alpha_5 - 1200\alpha_1'\alpha_5 - 1200\alpha_1\alpha_5'$$
$$+ 1200\alpha_1'\alpha_5' + 1200\alpha_2\alpha_3 - 1200\alpha_2'\alpha_3 - 1200\alpha_2\alpha_3' + 1200\alpha_2'\alpha_3'$$
$$+ 1500\alpha_2\alpha_4 - 1500\alpha_2'\alpha_4 - 1500\alpha_2\alpha_4' + 1500\alpha_2'\alpha_4' + 1800\alpha_2\alpha_5$$
$$- 1800\alpha_2'\alpha_5 - 1800\alpha_2\alpha_5' + 1800\alpha_2'\alpha_5' + 2000\alpha_3\alpha_4 - 2000\alpha_3'\alpha_4$$
$$- 2000\alpha_3\alpha_4' + 2000\alpha_3'\alpha_4' + 2400\alpha_3\alpha_5 - 2400\alpha_3'\alpha_5 - 2400\alpha_3\alpha_5'$$
$$+ 2400\alpha_3'\alpha_5' + 3000\alpha_4\alpha_5 - 3000\alpha_4'\alpha_5 - 3000\alpha_4\alpha_5' + 3000\alpha_4'\alpha_5'\Big\}\Big]$$

Subject to

$$\alpha_1 + \alpha_2 + \alpha_3 + \alpha_4 + \alpha_5 = \alpha_1' + \alpha_2' + \alpha_3' + \alpha_4' + \alpha_5'$$

$$\alpha_i \geq 0,$$

$$\alpha_i' \geq 0, \quad i = 1, 2, ..., l.$$

The optimum values of the Lagrange multipliers are given by

$$\alpha_1^* = 0.28$$

$$\alpha_1'^* = 0$$

$$\alpha_2^* = 0$$

$$\alpha_2'^* = 0.71$$

$$\alpha_3^* = 0$$

$$\alpha_3'^* = 0$$

$$\alpha_4^* = 1$$

$$\alpha_4'^* = 0$$

$$\alpha_5^* = 0$$

$$\alpha_5'^* = 0.57$$

Hence, the optimum value of the objective function $Q(\alpha)$ of the dual problem becomes

$$Q(\alpha) = 2.4 \times 0.28 - 3.7 \times 0.71 + 6.9 \times 1 - 7.6 \times 0.57 - \frac{1}{2}\Big[400 \times 0.28^2$$
$$+ 900 \times 0.71^2 + 2500 \times 1^2 + 3600 \times 0.57^2 + 2\{-600 \times 0.28 \times 0.71$$
$$+ 1000 \times 0.28 \times 1 - 1200 \times 0.28 \times 0.57 - 1500 \times 0.71 \times 1 + 1800$$
$$\times 0.71 \times 0.57 - 3000 \times 1 \times 0.57\}\Big]$$

$$= 0.608$$

From Eq. (3.48), we find the optimum value of weight is

$$w^* = \sum_{i=1}^{l}\left(\alpha_i^* - \alpha_i'^*\right)x_i$$

$$= (0.28 - 0)x_1 + (0 - 0.71)x_2 + (0 - 0)x_3 + (1 - 0)x_4 + (0 - 0.57)x_5$$

$$= 0.28x_1 - 0.71x_2 + x_4 - 0.57x_5$$

$$= 0.28 \times 20 - 0.71 \times 30 + 50 - 0.57 \times 60$$

$$= 0.13$$

Now, from Eq. (3.50), we have

$$b^* = \begin{cases} y_i - w^{*T}x - \epsilon, & \text{for } 0 < \alpha_i^* < C \\ y_i - w^{*T}x + \epsilon, & \text{for } 0 < \alpha_i^* < C \end{cases}$$

Assuming $C=1$, we have
$\alpha_1^* = 0.28$, for $0 < \alpha_i^* < C$
Hence,

$$b^* = y_i - w^{*T}x - \epsilon$$
$$= 2.5 - 0.13 \times 20 - 0.1$$
$$= -0.2$$

Therefore, the optimal regression line becomes

$$\hat{y} = w^{*T}x + b^*$$
$$= 0.13x - 0.2$$

The lines that constitute the tubes are

$$\hat{y} + \epsilon = w^{*T}x + b^* + \epsilon$$
$$= 0.13x - 0.2 + 0.1$$
$$= 0.13x - 0.1$$

and

$$\hat{y} - \epsilon = w^{*T}x + b^* - \epsilon$$
$$= 0.13x - 0.2 - 0.1$$
$$= 0.13x - 0.3$$

Fig. 3.19 depicts the optimum regression line with the tube.

3.3.2 Application of SVR in prediction of silk renditta

The raw silk renditta is an important quality parameter of silk which is defined as the quantity of cocoons in kg required to produce 1 kg of raw silk. Naturally, the requirement of a higher quantity of cocoon means the quality of silk is inferior and vice versa. SVR was employed to predict raw silk renditta from four parameters derived from silk cocoons. These are defective cocoon (%), shell ratio (%), number of cocoons/L, and cocoon mass. Table 3.7 shows the measured values of the cocoon parameters and corresponding raw silk renditta of 43 different lots of mulberry silk cocoons.

A sequence of arithmetic progress with a common difference of 5, such as 5th, 10th, 15th, ..., 40th data points, is used as a testing dataset, and remaining 35 data points are used as a training dataset. The Gaussian radial basis kernel function has been used for data mapping. The percentage accuracy of prediction

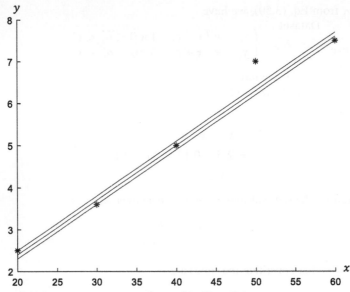

FIG. 3.19 Linear support vector regression model with $\epsilon = 0.1$.

TABLE 3.7 Dataset of cocoon parameters and renditta.

Defective cocoons (%)	Shell ratio (%)	No. of cocoons/L	Mass of 100 cocoons (g)	Renditta
9.9	16.8	122	138	12.1
9.3	15.3	143	141	10.8
7.4	14.7	142	138	11
8.2	14.2	139	141	11.2
8.5	15	141	139	11.1
9.8	14.9	140	143	11.1
8.9	15	138	142	11
8.4	15.2	141	144	10.4
11	15.7	132	191	10.8
10.8	17.2	129	194	10.9
11.1	17.7	131	195	10.9
10.4	18	134	182	10.9
10.1	17.3	131	192	11.6

TABLE 3.7 Dataset of cocoon parameters and renditta—cont'd

Defective cocoons (%)	Shell ratio (%)	No. of cocoons/L	Mass of 100 cocoons (g)	Renditta
10.7	17.2	129	174	11.4
10.5	17.6	137	181	11.2
11.8	17.9	141	178	11.1
10.5	18	135	173	11
10.6	17.2	142	182	11.4
11.7	17.1	136	179	11.4
11.8	17.6	139	171	11
10.4	17.7	139	168	11.2
11	17.8	141	172	10.9
11.1	17.4	144	165	10.4
11	17.8	138	172	11.9
11.8	17.9	137	169	12.1
11.1	16.2	128	183	11.9
12.4	17.3	136	191	11.6
12.4	16.5	145	179	11.4
9.8	15.6	132	167	10.8
9.9	15.9	141	178	10.6
10.1	15.7	132	194	10.8
10.1	15	139	186	10.8
9.8	15.9	128	191	10.6
9.9	15.8	127	195	10.4
9.1	15.1	126	180	10
8.5	16	155	145	10.3
11.4	15.7	136	184	10.9
11	16.1	127	167	10.8
10.8	15.4	142	176	10.7
9.6	15.8	142	168	10.6
7.5	16.7	136	154	10.9
8.2	16.4	138	159	10.6
8.2	18	148	153	11.8

for training and testing datasets is depicted in Tables 3.8 and 3.9, respectively. The mean absolute percentage accuracy between the actual and predicted values for training and testing sets is estimated as 97.82% and 97.10%, respectively. A MATLAB® coding was used to execute the computational work.

TABLE 3.8 Training performances.

Actual value	Predicted value	Accuracy (%)
12.1	11.39	94.10
10.8	10.84	99.59
11	10.96	99.60
11.2	11.12	99.25
11.1	10.99	99.03
11	10.99	99.93
10.4	10.85	95.71
10.8	10.84	99.59
10.9	11.09	98.25
10.9	11.03	98.81
11.6	10.97	94.53
11.4	11.10	97.38
11.1	11.14	99.60
11	11.04	99.60
11.4	11.06	97.05
11.4	11.23	98.53
11.2	11.01	98.32
10.9	11.05	98.66
10.4	10.99	94.28
11.9	11.08	93.08
11.9	10.94	91.92
11.6	11.34	97.74
11.4	11.23	98.53
10.8	10.80	99.98

TABLE 3.8 Training performances—cont'd

Actual value	Predicted value	Accuracy (%)
10.8	10.74	99.42
10.8	10.76	99.59
10.6	10.70	99.08
10.4	10.68	97.28
10.3	10.61	96.97
10.9	10.96	99.48
10.8	11.02	98.01
10.7	10.84	98.69
10.9	10.86	99.59
10.6	10.77	98.42
11.8	11.09	94.01
Mean accuracy (%) of training set		97.82

TABLE 3.9 Testing performances.

Actual value	Predicted value	Accuracy (%)
11.1	10.93	98.47
10.9	11.02	98.90
11.2	11.06	98.75
11.0	11.16	98.50
12.1	11.15	92.12
10.6	10.80	98.16
10.0	10.69	93.06
10.6	10.73	98.82
Mean accuracy (%) of testing set		97.10

3.4 MATLAB coding

3.4.1 MATLAB coding of SVC example given in Section 3.2.1

```
clc
clear
close all
format short
x=[0.5 0.5
    0.3 0.7
    0.1 0.1
    0.2 0];
y=[1
    1
    -1
    -1];
K=x*x';
fori=1:length(y)
for j=1:length(y)
    Y(i,j)=[y(i)*y(j)];
end
end
H=(Y.*K);
f=ones(length(y),1);
Aeq = y';
beq = 0;
A=[-1 0 0 0
    0 -1 0 0
    0 0 -1 0
    0 0 0 -1];
B=[0 0 0 0];
[alpha,fval]=quadprog(H,-f,A,B,Aeq,beq);
[s1,s2]=size(x);
x=x';
w=zeros(s2,1);
fori=1:length(y)
    w=w+alpha(i)*y(i)*x(:,i);
end
w
x
y1=find(y==1)
y2=find(y==-1)
s=w'*x
b=-(max(s(y2))+min(s(y1)))/2
```

3.4.2 MATLAB coding of SVC example given in Section 3.2.2

```
clc
clear
close all
format short
x=[1       1
    1      -1
   -1      -1
   -1       1
    0.5    0.5
  0.5   -0.5
  -0.5   -0.5
  -0.5    0.5];
y=[1
   1
   1
   1
   -1
   -1
   -1
   -1];
fori=1:length(y)
phi_x(i,:)=[x(i,1)^2 x(i,2)^2 sqrt(2)*x(i,1)*x(i,2)];
end
phix=phi_x';
K=phix'*phix;
fori=1:length(y)
for j=1:length(y)
    Y(i,j)=[y(i)*y(j)];
end
end
Y;
H=(Y.*K);
f=ones(length(y),1);
Aeq = y';
beq = 0;
A=[-1 0 0 0 0 0 0 0
    0 -1 0 0 0 0 0 0
    0 0 -1 0 0 0 0 0
    0 0 0 -1 0 0 0 0
    0 0 0 0 -1 0 0 0
    0 0 0 0 0 -1 0 0
    0 0 0 0 0 0 -1 0
    0 0 0 0 0 0 0 -1];
```

```
B=[0 0 0 0 0 0 0 0];
[alpha,fval]=quadprog(H,-f,A,B,Aeq,beq)
[s1,s2]=size(phix');
w=zeros(s2,1);
fori=1:length(y)
    w=w+alpha(i)*y(i)*phix(:,i);
end
w
y1=find(y==1);
y2=find(y==-1);
s=w'*phix
b=-(max(s(y2))+min(s(y1)))/2
```

3.4.3   MATLAB coding of SVC application given in Section 3.2.3

```
clc
clear
close all
data= [18.635    54.008    1
        17.332    83.395    1
        19.302    56.474    1
        17.713    38.898    1
        14.397    55.843    1
        12.726    57.288    1
        51.872    44.926    1
        30.331    65.963    1
        29.348    67.969    1
        27.231    50.54     1
        22.611    45.371    1
        27.696    44.358    1
        14.618    50.691    1
        40.119    59.768    1
        23.064    44.229    1
        10.59     46.401    1
        51.293    59.983    1
        39.611    49.976    1
        16.771    50.425    1
        35.978    44.431    1
        30.615    53.942    1
        17.096    44.415    1
        24.487    66.153    1
        26.823    36.617    1
        25.505    50.805    1
```

```
         11.942    9.8894   -1
         7.6811    7.8098   -1
         11.487    9.0584   -1
         13.479    5.7755   -1
         11.303    14.093   -1
         11.606    17.433   -1
         14.937    13.507   -1
         18.44     17.199   -1
         9.6977    8.9249   -1
         18.345    11.234   -1
         7.3993    10.518   -1
         21.398    18.8     -1
         25.536    10.706   -1
         23.929    13.37    -1
         16.476    14.566   -1
         14.821    15.221   -1
         11.708    13.201   -1
         11.023    16.858    1
         15.263    14.557   -1
         8.5294    11.236   -1
         10.712    11.41    -1
         9.7852    13.458   -1
         9.7468    16.161   -1
         12.994    12.045   -1
         12.587    17.842   -1];
[s1,s2]=size(data);
n=s1;
for j=1:2
    norm_data(:,j)=(data(:,j)-min(data(:,j)))/(max(data(:,j))-
min(data(:,j)));
end
input_array=norm_data;
target_array=data(:,end);
k=5;
p=0;
q=0;
r=0;
s=1:n;
fori=1:n
if rem(i,k)==0
      p=p+k;
      q=q+1;
tst(q)=s(p);
```

```
else
        r=r+1;
trn(r)=s(i);
end
end
input_array_training=input_array(trn,:);
target_array_training=target_array(trn);
input_array_testing=input_array(tst,:);
target_array_testing=target_array(tst);
mdl = fitcsvm(input_array_training,target_array_training,
'Standardize',false,...
'kernelfunction','linear')
predicted_value_testing=predict(mdl,input_array_testing);
predicted_value_training=predict(mdl,input_array_training);
y=target_array_testing;
y1=predicted_value_testing;
accuracy_testing=mean(100-((abs(y-y1)./y)*100))
z=target_array_training;
z1=predicted_value_training;
mean_accuracy_training=mean(100-((abs(z-z1)./z)*100))
sv = mdl.SupportVectors;
figure,gscatter(input_array_training(:,1),input_array_training
(:,2),target_array_training,'kk','*x')
hold on
plot(sv(:,1),sv(:,2),'ko','MarkerSize',10)
hold on
x1=input_array(:,1);
x2=input_array(:,2);
h =0.01; % Mesh grid step size
[X1,X2] = meshgrid(-1.1:h:1.1,-1.1:h:1.1);
[~,score] = predict(mdl,[X1(:),X2(:)]);
scoreGrid=reshape(score(:,1),size(X1,1),size(X1,2));
contour(X1,X2,scoreGrid,[1 0 -1],'k')
legend('Power-loom fabrics','Hand-loom fabrics','Support Vector')
xlim([-0.1 1.1])
ylim([-0.1 1.1])
xticks([])
yticks([])
set(gcf,'color','w')
```

3.4.4   MATLAB coding of SVC application given in Section 3.2.4

```
clc
clear
close all
%
defect=[0.3978   0.6433   0.3704   0.443    0.3584   0.3811   1
        0.392    0.6464   0.3532   0.4221   0.3352   0.3859   1
        0.3887   0.6363   0.3601   0.4202   0.322    0.3257   1
        0.388    0.6322   0.3672   0.4302   0.3481   0.3378   1
        0.3851   0.6228   0.3567   0.4361   0.3496   0.3371   1
        0.39     0.6402   0.3584   0.4205   0.3726   0.3434   1
        0.4026   0.6362   0.3601   0.432    0.3438   0.3442   1
        0.3879   0.6161   0.3419   0.4153   0.3228   0.3547   1
        0.3931   0.6381   0.3569   0.4284   0.3694   0.4308   1
        0.3826   0.6298   0.3537   0.4234   0.3489   0.3435   1
        0.3529   0.5768   0.3219   0.3865   0.4417   0.4725   2
        0.3465   0.5874   0.3225   0.3819   0.474    0.5255   2
        0.3467   0.5767   0.313    0.3782   0.3845   0.4925   2
        0.3697   0.5805   0.3232   0.3978   0.466    0.4953   2
        0.3537   0.5642   0.3182   0.3918   0.4358   0.5035   2
        0.3689   0.6188   0.3483   0.4026   0.4393   0.4813   2
        0.3789   0.6173   0.3447   0.4042   0.3954   0.4213   2
        0.3663   0.6173   0.3444   0.4045   0.4439   0.4788   2
        0.3881   0.6345   0.3569   0.4305   0.4214   0.5121   2
        0.3964   0.6362   0.3512   0.4236   0.4049   0.421    2
        0.3159   0.5158   0.3214   0.3981   0.5433   0.3301   3
        0.3354   0.5356   0.3373   0.4095   0.5594   0.3677   3
        0.3231   0.5202   0.3197   0.3899   0.5466   0.351    3
        0.3534   0.5655   0.3275   0.4129   0.521    0.3302   3
        0.3761   0.5795   0.3399   0.4324   0.529    0.3305   3
        0.3509   0.5957   0.3507   0.4079   0.5432   0.3107   3
        0.3661   0.5915   0.3361   0.4137   0.4808   0.2884   3
        0.3717   0.5968   0.3237   0.4003   0.4708   0.3376   3
        0.3589   0.5903   0.323    0.3931   0.4377   0.3266   3
        0.3436   0.5775   0.3298   0.3907   0.4888   0.3454   3
        0.3765   0.608    0.3098   0.3842   0.3198   0.3587   4
        0.3987   0.6132   0.3145   0.3954   0.3272   0.3829   4
        0.384    0.5953   0.3123   0.392    0.3165   0.4022   4
        0.3854   0.6023   0.3101   0.389    0.3154   0.3635   4
        0.3873   0.597    0.3074   0.3944   0.3554   0.3735   4
        0.3723   0.5821   0.2097   0.3695   0.3453   0.3765   4
        0.3836   0.6022   0.3054   0.3861   0.3383   0.3429   4
        0.3716   0.5918   0.3101   0.3761   0.3595   0.3248   4
```

```
     0.4115   0.6037   0.2797   0.4036   0.3987   0.3294   4
     0.4321   0.6446   0.309    0.4157   0.4254   0.3284   4
     0.3592   0.4453   0.3003   0.3543   0.4673   0.41     5
     0.4049   0.4874   0.3207   0.3977   0.5187   0.424    5
     0.3586   0.4805   0.3102   0.3614   0.4967   0.8066   5
     0.3049   0.3866   0.2726   0.3215   0.4967   0.5492   5
     0.4029   0.5257   0.3363   0.4028   0.5465   0.4661   5
     0.4      0.4976   0.3254   0.3969   0.5242   0.4233   5
     0.2626   0.3115   0.2417   0.2633   0.4584   0.3841   5
     0.2657   0.3276   0.2263   0.2723   0.3681   0.4321   5
     0.364    0.4823   0.3034   0.3518   0.5274   0.62     5
     0.4051   0.5158   0.3361   0.4082   0.6228   0.6095   5];
[s1,s2]=size(defect);
random_sample=randperm(s1);%Data randomisation
defect=defect(random_sample,:);
%Input data normalisation
for j=1:6
    norm_data(:,j)=(defect(:,j)-min(defect(:,j)))/(max(defect
(:,j))-min(defect(:,j)));
end
n=s1;%Number of experiments or data points
input_array=norm_data;
target_array=defect(:,end);
% Division of training and testing arrays of input pattern
Y=[];
Y1=[];
Z=[];
Z1=[];
for i=1:n
input_array_testing=input_array(i,:);
target_array_testing=target_array(i);
if i==1
input_array_training=input_array(i+1:end,:);
target_array_training=target_array(i+1:end);
elseif i==n
input_array_training=input_array(1:n-1,:);
target_array_training=target_array(1:n-1);
else
input_array_training=input_array([1:(i-1) (i+1):end],:);
target_array_training=target_array([1:(i-1) (i+1):end]);
end
% Support vector classification
% training
```

```
t = templateSVM('KernelFunction','gaussian','KernelScale','auto');
mdl = fitcecoc(input_array_training,target_array_training,'Lear-
ners',t);
%Predicted output of testing and training
predicted_value_testing=predict(mdl,input_array_testing);
predicted_value_training=predict(mdl,input_array_training);
%Prediction statistics
    y=target_array_testing;
    y1=predicted_value_testing;
accuracy_testing(i)=100-((abs(y-y1)./y)*100);
    Y=[Y y'];
    Y1=[Y1 y1'];
    z=target_array_training;
    z1=predicted_value_training;
mean_accuracy_training(i)=mean(100-((abs(z-z1)./z)*100));
    Z=[Z z'];
    Z1=[Z1 z1'];
end
%Prediction results
Testing_accuracy=accuracy_testing';
Training_accuracy=mean_accuracy_training';
grand_mean_accuracy_testing=mean(100-((abs(Y-Y1)./Y)*100))
grand_mean_accuracy_training=mean(100-((abs(Z-Z1)./Z)*100))
```

3.4.5   MATLAB coding of SVR example given in Section 3.3.1

```
clc
clear
close all
format bank
data=[20   2.5
30   3.6
40   5
50   7
60   7.5];
x=data(:,1)';
y=data(:,2)';
n=length(y);
j=1;
for i=1:n
  X(1,[j,j+1])=[x(i) -x(i)];
  j=j+2;
end
```

```
H=X'*X;
epsilon=0.1;
j=1;
for i=1:n
   Y(1,[j,j+1])=[(y(i)-epsilon) (-y(i)-epsilon)];
   j=j+2;
end
f=Y;
e=ones(length(y),1);
j=1;
for i=1:n
   E(1,[j,j+1])=[e(i) -e(i)];
   j=j+2;
end
Aeq = E;
beq = 0;
C=1;
lb=zeros(1,2*n);
ub=C*ones(1,2*n);
[alpha,fval]=quadprog(H,-f,[],[],Aeq,beq,lb,ub)
j=1;
fori=1:n
ALPHA(i,:)=alpha(j)-alpha(j+1);
alph(i)=alpha(j);
alph_1(i)=alpha(j+1);
j=j+2;
end
ALPHA
w=sum(ALPHA.*x')
j=1;k=1;
tol=0.000001;
fori=1:n
if (alph(i)>tol)&&(alph(i)<(1-tol)*C)
b1(j)=y(i)-w*x(i)-epsilon;
j=j+1;
end
if (alph_1(i)>tol)&&(alph_1(i)<(1-tol)*C)
b2(k)=y(i)-w*x(i)+epsilon;
      k=k+1;
end
end
b1
b2
```

3.4.6   MATLAB coding of SVR application given in Section 3.3.2

```
clc
clear
close all
%
silk_data=[9.9    16.8    122    138    12.1
           9.3    15.3    143    141    10.8
           7.4    14.7    142    138    11.0
           8.2    14.2    139    141    11.2
           8.5    15.0    141    139    11.1
           9.8    14.9    140    143    11.1
           8.9    15.0    138    142    11.0
           8.4    15.2    141    144    10.4
          11.0    15.7    132    191    10.8
          10.8    17.2    129    194    10.9
          11.1    17.7    131    195    10.9
          10.4    18.0    134    182    10.9
          10.1    17.3    131    192    11.6
          10.7    17.2    129    174    11.4
          10.5    17.6    137    181    11.2
          11.8    17.9    141    178    11.1
          10.5    18.0    135    173    11.0
          10.6    17.2    142    182    11.4
          11.7    17.1    136    179    11.4
          11.8    17.6    139    171    11.0
          10.4    17.7    139    168    11.2
          11.0    17.8    141    172    10.9
          11.1    17.4    144    165    10.4
          11.0    17.8    138    172    11.9
          11.8    17.9    137    169    12.1
          11.1    16.2    128    183    11.9
          12.4    17.3    136    191    11.6
          12.4    16.5    145    179    11.4
           9.8    15.6    132    167    10.8
           9.9    15.9    141    178    10.6
          10.1    15.7    132    194    10.8
          10.1    15.0    139    186    10.8
           9.8    15.9    128    191    10.6
           9.9    15.8    127    195    10.4
           9.1    15.1    126    180    10.0
           8.5    16.0    155    145    10.3
          11.4    15.7    136    184    10.9
          11.0    16.1    127    167    10.8
```

```
          10.8   15.4   142   176   10.7
           9.6   15.8   142   168   10.6
           7.5   16.7   136   154   10.9
           8.2   16.4   138   159   10.6
           8.2   18.0   148   153   11.8];
[s1,s2]=size(silk_data);
%Input data normalisation
for j=1:4
    norm_data(:,j)=(silk_data(:,j)-min(silk_data(:,j)))/(max
(silk_data(:,j))-min(silk_data(:,j)));
end
norm_data;
n=s1;%Number of experiments or data points
input_array=norm_data;
target_array=silk_data(:,end);
% Division of training and testing arrays of input pattern
k=5;%common difference
p=0;
q=0;
r=0;
s=1:n;
for i=1:n
if rem(i,k)==0
        p=p+k;
        q=q+1;
tst(q)=s(p);
else
        r=r+1;
trn(r)=s(i);
end
end
tst;
trn;
input_array_training=input_array(trn,:);
target_array_training=target_array(trn);
input_array_testing=input_array(tst,:);
target_array_testing=target_array(tst);
% train SVR
mdl =
fitrsvm(input_array_training,target_array_training,'Standardize',
true,...
'kernelfunction','gaussian','KernelScale','auto');
%Predicted output of testing and training
```

```
predicted_value_testing=predict(mdl,input_array_testing);
predicted_value_training=predict(mdl,input_array_training);
%Prediction statistics
y=target_array_testing;
y1=predicted_value_testing;
accuracy_testing=mean(100-((abs(y-y1)./y)*100))
z=target_array_training;
z1=predicted_value_training;
mean_accuracy_training=mean(100-((abs(z-z1)./z)*100))
```

3.5 Summary

This chapter on SVM provides an overview of this powerful machine learning algorithm that is usually used for classification and regression tasks. SVM uses kernel trick which transforms the original data in a higher dimension feature space where it finds the linear optimal separating hyperplane or decision boundary. The chapter begins by introducing the concept of SVM and its basic principles. It explains the key concepts of SVC, including the notion of support vectors, which are the data points closest to the decision boundary. It discusses in detail the optimization problem of maximizing the margin associated with SVC. The use of different kernel functions, such as linear, polynomial, and Gaussian radial basis functions, is also explained. Step-by-step examples of SVC for the classification of both linearly and nonlinearly separable data are discussed. It then delves into the concept of SVR with step-by-step examples. The authors also shed light on the application of SVC and SVR in the domain of textile engineering with their findings. This chapter serves as a solid foundation for understanding and exploring the potential of SVM in solving complex problems of classification and regression across different domains.

References

Cristianini, N., & Shawe-Taylor, J. (2000). *An introduction to support vector machines and other kernel based learning methods*. Cambridge: Cambridge University Press.

Fung, G. M., & Mangasarian, O. L. (2005). Multi-category proximal support vector machine classifiers. *Machine Learning, 59*, 77–97.

Ghosh, A., Guha, T., & Bhar, R. B. (2015). Identification of handloom and powerloom fabrics using proximal support vector machines. *Indian Journal of Fibre & Textile Research, 40*, 87–93.

Ghosh, A., Guha, T., Bhar, R. B., & Das, S. (2011). Pattern classification of fabric defects using support vector machines. *International Journal of Clothing Science and Technology, 23*, 142–158.

Hamel, L. (2009). *Knowledge discovery with support vector machines*. Hoboken, NJ: John Wiley & Sons, Inc.

Han, J., & Kamber, M. (2006). *Data mining: Concepts and techniques* (2nd ed.). Amsterdam: Elsevier.

Schölkopf, B., & Smola, A. J. (2002). *Learning with kernels: Support vector machines, regularization, optimization, and beyond.* USA: MIT Press.

Suykens, J. A. K., Gestel, T. D., Brabanter, J., de Moor, B., & Vandewalle, J. (2002). *Least-square support vector machines.* Singapore: World Scientific Publishing Co. Pvt. Ltd.

Tsai, I., Lin, C., & Lin, J. (1995). Applying an artificial neural network to pattern recognition in fabric defects. *Textile Research Journal, 65,* 123–130.

Vapnik, V. N., & Chervonenkis, A. (1964). A note on one class of perceptrons. *Automation and Remote Control, 25,* 103–109.

Vapnik, V. N. (1995). *The nature of statistical learning theory.* New York: Springer Verlag.

Chapter 4

Fuzzy logic

4.1 Introduction

Fuzzy logic is a method of reasoning, which imitates the human reasoning system. The approach resembles the human decision-making process that involves all intermediate possibilities between yes or no and true or false. For example, if someone asks a question about a glass of cold water: "Is it cold"? There are two possibilities in Boolean logic, Yes or No. But in fuzzy logic, the answer could be very much, little, not much, very little, etc. The inventor of fuzzy logic, Prof L. A. Zadeh (Zadeh, 1965) of the University of California, observed that, unlike computers, humans have different ranges of possibilities between Yes/1 and No/0 such as "Certainly Yes," "Possibly Yes," "Can't Say," "Certainly No," and "Possibly No". In fuzzy logic, a definite output can be obtained from different possibilities of input. In general, real-world problems are associated with different types of uncertainties and imprecision. Before 1965, researchers used to consider classical probability theory, which is based on crisp logic or bivalent logic, as the prime tool for dealing with uncertainties. Prof. Zadeh was the first to argue that some uncertainties cannot be tackled using the probability theory. For a simple example, a little daughter asks her father to bring a blue dress for her from the market. There are two uncertainties at least, (i) availability of the baby dress and (ii) a guarantee that the dress is blue. There is a probability of obtaining a baby dress (i.e., the frequency of likelihood that an element is in a class), which varies between 0 and 1. According to the crisp set, the baby dress will be either blue (1) or non-blue (0). On the contrary, in a fuzzy set, the color blue can be defined using the concept of membership (i.e., similarity of an element to a class) value. If the color is perfectly blue, then it may be said blue with a membership value of 1, if it is almost blue, then it is considered blue with a membership value of 0.9, if it is slightly blue, then it is assumed to be blue with a membership value of 0.2, even if it is not blue, then also it is called blue with a membership value of 0, and so on. Thus, in the fuzzy set, an element can be a member of the set with some membership value (i.e., degree of belongingness). In this way, the uncertainty related to the color of the baby dress can be handled using fuzzy logic.

As fuzzy logic is tolerant to imprecision, uncertainty, partial truth, and approximation, and it is based on human reasoning, it has empowered scientists and technologists from diverse engineering disciplines. Fuzzy logic provides

Artificial Intelligence in Textile Engineering. https://doi.org/10.1016/B978-0-443-15395-2.00003-X

flexibility and reasoning in artificial intelligence. It is also an appealing tool for textile product engineering as it can handle imprecision that is present in the textile data. For example, a spinner often uses terms such as "fine" and "coarse" to assess the fiber and yarn count, although these terms do not constitute a well-defined boundary. In textiles, the rapidly bourgeoning sway of fuzzy logic is now well recognized in the selection of raw material, setting and control of process parameters, classification of patterns, and prediction of the properties of various fibrous products. In this chapter, implementation of fuzzy logic in AI is explicated in detail.

4.2   Concept of crisp set and fuzzy set

In a classical sense, the collection of well-defined distinct objects is known as a set. For example, a set of natural numbers or a set of integers. The elements belonging to the set of natural numbers possess some features by which one can identify them as natural numbers. That feature is the surrogate of well-definedness. On the contrary, a collection of short people or a collection of tall people is not a set unless the shortness or tallness is well-defined. To have a better understanding, suppose we call a person short if its height is less than 5 ft. 6 in. Now a person having a height of 5 ft. 5.8 in is not that short as compared to a person having a height of 5 ft. The degree of shortness is not the same for both the persons. In the classical sense, both the persons belong to the same set, i.e., set of short people although the heights of both the persons differ significantly. To avoid this discrepancy, we can assign some degree of shortness to distinguish the persons that is done by suitably defining membership function, which represents the degree of eligibility to belong to that set. Thus, a fuzzy set can be defined in the following way:

Definition 4.1 A fuzzy set $A(x)$ on a set X is a collection of ordered tuples as follows.

$$A(x) = \{(x, \mu_A(x)) | x \in X\}$$

where $\mu_A : X \rightarrow [0, 1]$ is called the membership function of the set $A(x)$.

4.3   Membership functions

The fuzziness of the set is defined by the membership functions irrespective of the elements in the set, which are discrete or continuous. The membership functions are used as a technique to solve the empirical problem by experience rather than knowledge (Zimmerman, 1996). It is developed based on the opinion of experts to solve the practical problems. The membership functions are generally represented in graphical form. A fuzzy set $\tilde{A}$ in the universe of information

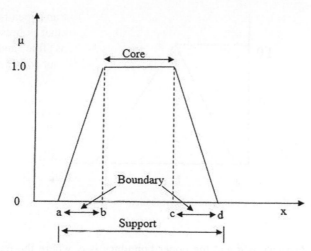

FIG. 4.1 Features of the membership function.

U can be defined as a set of ordered pairs and it can be represented mathematically as

$$\tilde{A} = \{x, \mu_{\tilde{A}}(x) | x \in U\} \tag{4.1}$$

where $\mu_{\tilde{A}}(\bullet)$ is membership function of $\tilde{A}$. The membership function $\mu_{\tilde{A}}(\bullet)$ maps to U to the membership space M, i.e., $\mu_{\tilde{A}} : X \to M$. The membership value ranges from 0 to 1. The dot ($\bullet$) in the membership function represents the element in a fuzzy set, whether it is discrete or continuous.

The three basic features of the membership function are shown in Fig. 4.1.

i. *Core:* The core of the membership function is that region (bc) of the universe that is characterized by full membership in the set. Hence, the core consists of all those elements of the universe of information such that, $\mu_{\tilde{A}}(x) = 1$

ii. Support: The support of a membership function is the region (ad) of the universe that is characterized by a non-zero membership in the set. Hence core consists of all those elements of the universe of information such that, $\mu_{\tilde{A}}(x) > 0$

iii. Boundary: The boundary of a membership function is the region (ab and cd) of the universe that is characterized by a nonzero but incomplete membership in the set. Hence, the core consists of all those elements of the universe of information such that, $0 < \mu_{\tilde{A}}(x) < 1$

4.3.1 Triangular membership function

Fig. 4.2 shows the triangular membership function distribution. Let a, b, and c represent the x coordinates of the three vertices of $\mu_{\tilde{A}}(x)$ in a fuzzy set

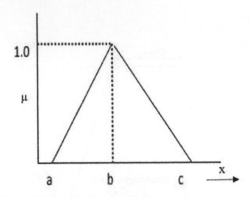

FIG. 4.2 Triangular membership function distribution.

A. Lower boundary is a and the upper boundary is c, where the membership degree is zero. The b is the center where the membership degree is 1.

$$\text{Triangle } (x; a, b, c) = 0 \quad \text{if } x \leq a$$

$$= \frac{x - a}{b - a} \quad \text{if } a \leq x \leq b$$

$$= \frac{c - x}{c - b} \quad \text{if } b \leq x \leq c \qquad (4.2)$$

$$= 0 \quad \text{if } x \geq c$$

The μ_{triangle} can be expressed as follows:

$$\mu_{\text{triangle}} = \max\left(\min\left(\frac{x - a}{b - a}, \frac{c - x}{c - b}\right), 0\right) \qquad (4.3)$$

The membership function values at $x = a$, $x = b$, and $x = c$ are set equal to 0, 1, and 0, respectively.

4.3.2 Trapezoidal membership function

Let a, b, c, and d represent the x coordinates of the membership function. Fig. 4.3 represents the trapezoidal membership function distribution

$$\text{Trapizoid } (x; a, b, c, d) = 0 \quad \text{if } x \leq a$$

$$= \frac{x - a}{b - a} \quad \text{if } a \leq x \leq b$$

$$= 1 \quad \text{if } b \leq x \leq c$$

$$= \frac{d - x}{d - c} \quad \text{if } c \leq x \leq d \qquad (4.4)$$

$$= 0 \quad \text{if } x \geq d$$

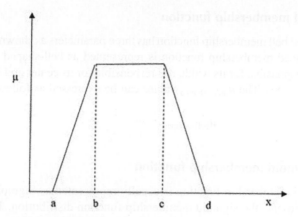

FIG. 4.3 Trapezoidal membership function distribution.

The membership function values at $x=a$, $x=b$, $x=c$, and $x=d$ are set equal to 0, 1, 1, and 0, respectively. The $\mu_{\text{trapizoidal}}$ values can be expressed as follows:

$$\mu_{\text{trapizoidal}} = \max\left(\min\left(\frac{x-a}{b-a}, 1, \frac{d-x}{d-c}\right), 0 \right) \tag{4.5}$$

4.3.3 Gaussian membership function

The Gaussian membership function is usually represented as Gaussian $(x; m, \sigma)$ where m and σ represent the mean and standard deviations. Fig. 4.4 represents the Gaussian membership function distribution. The μ_{gaussian} values can be expressed as follows:

$$\mu_{\text{gaussian}} = \frac{1}{e^{\frac{1}{2}\left(\frac{x-m}{\sigma}\right)^2}} \tag{4.6}$$

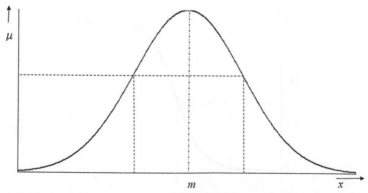

FIG. 4.4 Gaussian membership function distribution.

4.3.4 Bell membership function

A generalized bell membership function has three parameters as shown in Fig. 4.5. The bell-shaped membership function is represented as bell-shaped $(x; a, b, c)$ where a is responsible for its width, c is responsible for its center and b is responsible for its slopes. The $\mu_{\text{bell–shaped}}$ values can be expressed as follows:

$$\mu_{\text{bell–shaped}} = \frac{1}{1 + \left|\frac{x-c}{a}\right|^{2b}} \tag{4.7}$$

4.3.5 Sigmoid membership function

The sigmoid membership function is usually represented as Sigmoid $(x; a, b)$. Fig. 4.6 represents the sigmoid membership function distribution. The μ_{sigmoid} values can be expressed as follows:

$$\mu_{\text{sigmoid}} = \frac{1}{1 + e^{-a(x-b)}} \tag{4.8}$$

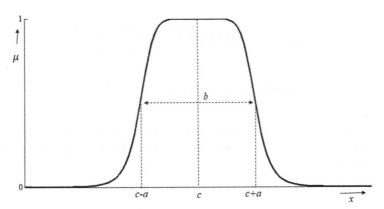

FIG. 4.5 Bell-shaped membership function distribution.

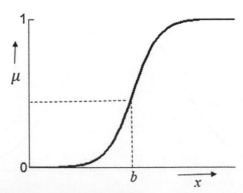

FIG. 4.6 Sigmoid membership function distribution.

where a is the slope of the distribution. The μ_{sigmoid} becomes 0.5 when $x=b$. It tends toward 0 and 1 as values of x approach 0 and higher value than b, respectively.

4.4 Basic operations on fuzzy sets

We are familiar with the basic operations such as union, intersection, and complementation on classical sets. In the following section, we discuss the similar basic operations on fuzzy sets (Sivanandam et al., 2010).

4.4.1 Union of sets

We know that for two crisp sets A and B, the union $A \cup B$ is defined as.

$$A \cup B = \{x | x \in A \text{ or } x \in B\} \tag{4.9}$$

On the other hand, if $(A(x), \mu_A)$ and $(B(x), \mu_B)$ are two fuzzy sets on a set X, the membership function of the union of the two given fuzzy sets is.

$$\mu_{A \cup B} = \max\{\mu_A, \mu_B\} \tag{4.10}$$

Thus, the union of the sets $(A(x), \mu_A)$ and $(B(x), \mu_B)$ is given by,

$$A(x) \cup B(x) = \{(x, \mu_{A \cup B}) | x \in X\}$$

Example 4.1 Let us consider the following two fuzzy sets.

$$A(x) = \{(x_1, 0.2), (x_2, 0.4), (x_3, 0.6), (x_4, 0.9)\}$$
$$B(x) = \{(x_1, 0.1), (x_2, 0.3), (x_3, 0.7), (x_4, 0.8)\}$$
$$\mu_{A \cup B}(x_1) = \max\{0.2, 0.1\} = 0.2$$
$$\mu_{A \cup B}(x_2) = \max\{0.4, 0.3\} = 0.4$$
$$\mu_{A \cup B}(x_3) = \max\{0.6, 0.7\} = 0.7$$
$$\mu_{A \cup B}(x_4) = \max\{0.9, 0.8\} = 0.9$$

Now according to the definition of union
$A(x) \cup B(x) = \{(x_1, 0.2), (x_2, 0.4), (x_3, 0.7), (x_4, 0.9)\}$.

4.4.2 Intersection of sets

For two crisp sets A and B, the intersection is defined as

$$A \cap B = \{x | x \in A \text{ and } x \in B\} \tag{4.11}$$

In the case of two fuzzy sets, $A(x)$ and $B(x)$ with membership functions μ_A and μ_B, respectively, the membership function of the intersection is

$$\mu_{A \cap B} = \min\{\mu_A, \mu_B\} \tag{4.12}$$

Therefore,

$$A(x) \cap B(x) = \{(x, \mu_{A \cap B}) | x \in X\}$$

Example 4.2 Let us consider the same fuzzy sets $A(x)$ and $B(x)$ as illustrated in Example 4.1. According to the definition of intersection

$$\mu_{A \cap B}(x_1) = \min\{(0.2, 0.1)\} = 0.1$$
$$\mu_{A \cap B}(x_2) = \min\{(0.4, 0.3)\} = 0.3$$
$$\mu_{A \cap B}(x_1) = \min\{(0.6, 0.7)\} = 0.6$$
$$\mu_{A \cap B}(x_1) = \min\{(0.9, 0.8)\} = 0.8$$

Therefore, $A(x) \cap B(x) = \{(x_1, 0.1), (x_2, 0.3), (x_3, 0.6), (x_4, 0.8)\}$.

4.4.3 Proper subset of a fuzzy set

Suppose $A(x)$ and $B(x)$ are two fuzzy sets having membership function μ_A and μ_B, respectively. $A(x) \subset B(x)$ if $\mu_A(x) < \mu_B(x)$ for all $x \in X$

Example 4.3 Suppose $A(x) = \{(x_1, 0.1), (x_2, 0.3), (x_3, 0.5), (x_4, 0.6)\}$ and $B(x) = \{(x_1, 0.2), (x_2, 0.4), (x_3, 0.6), (x_4, 0.7)\}$. Clearly, $_A(x) <_B(x)$ for all $x \in \{x_1, x_2, x_3, x_4\}$.

This shows that $A(x) \subset B(x)$

4.4.4 Equality of fuzzy set

Two fuzzy sets $A(x)$ and $B(x)$ with membership functions $\mu_A(x)$ and $\mu_B(x)$ are said to be equal if $\mu_A(x) = \mu_B(x)$ for all $x \in X$.

4.4.5 Complement of a fuzzy set

Suppose $A(x)$ is a fuzzy set with the membership function $\mu_A(x)$. Then the complement of $A(x)$ denoted by $\overline{A}(x)$ is the fuzzy set with the membership function $\mu_{\overline{A}}(x) = 1 - \mu_A(x)$ for all $x \in X$, that is

$$\overline{A}(x) = \{(x\mu_{\overline{A}}(x)) | x \in X\} \tag{4.13}$$

Example 4.4 Suppose that $A(x) = \{(x_1, 0.2), (x_2, 0.3), (x_3, 0.5), (x_4, 0.7)\}$. Then $\overline{A}(x) = \{(x_1 0.8), (x_2 0.7), (x_3 0.5), (x_4 0.3)\}$.

4.4.6 Multiplication of a fuzzy set by a crisp number

The product of fuzzy set $A(x)$ by a crisp number is given by

$$d.A(x) = \{(x, d.\mu_A(x)), x \in X\} \tag{4.14}$$

Example 4.5 Let us consider the fuzzy set $A(x) = \{(x_1, 0.1), (x_2, 0.4), (x_3, 0.5), (x_4, 0.7)\}$ and $d = 0.3$. Then $d. \, A(x) = \{(x_1, 0.03), (x_2, 0.12), (x_3, 0.15), (x_4, 0.21)\}$.

4.4.7 Sum of two fuzzy sets

If $A(x)$ and $B(x)$ are two fuzzy sets, then the algebraic sum of these two sets can be expressed as

$$A(x) + B(x) = \{(x, \mu_{A+B}(x)), x \in X\}, \tag{4.15}$$

where $\mu_{A+B}(x) = \mu_A(x) + \mu_B(x) - \mu_A(x). \, \mu_B(x)$.

Example 4.6 Let us consider the two fuzzy sets

$$A(x) = \{(x_1, 0.1), (x_2, 0.4), (x_3, 0.5), (x_4, 0.7)\}$$

and

$$B(x) = \{(x_1, 0.2), (x_2, 0.3), (x_3, 0.4), (x_4, 0.5)\}$$

So, according to the above definition

$$A(x) + B(x) = \{(x_1, 0.28), (x_2, 0.58), (x_3, 0.7), (x_4, 0.85)\}$$

4.4.8 Algebraic difference of fuzzy sets

Let $A(x)$ and $B(x)$ are two fuzzy sets having membership functions μ_A and μ_B, respectively. Consider the membership function

$$\mu_{A-B}(x) = \mu_{A \cap \overline{B}}.$$

The algebraic difference of the fuzzy sets $A(x)$ and $B(x)$ is the fuzzy set

$$A(x) - B(x) = \{(x, \mu_{A-B}(x)), x \in X\} \tag{4.16}$$

Example 4.7 To illustrate we consider the following two fuzzy sets

$$A(x) = \{(x_1, 0.1), (x_2, 0.6), (x_3, 0.8), (x_4, 0.9)\}$$

and

$$B(x) = \{(x_1, 0.2), (x_2, 0.5), (x_3, 0.8), (x_4, 0.9)\}$$
$$\overline{B}(x) = \{(x_1, 0.8), (x_2, 0.5), (x_3, 0.2), (x_4, 0.1)\}$$

Thus,

$$A(x) - B(x) = \{(x, \mu_{A-B}(x)), x \in X\} = \{(x, \mu_{A \cap \overline{B}}(x)), x \in X\}$$
$$= \{(x_1, 0.1), (x_2, 0.5), (x_3, 0.2), (x_4, 0.1)\}$$

4.5 Alpha cut of the fuzzy set

It is a set consisting of elements x of the Universal set X, whose membership values are either greater than or equal to the value α. The crisp set is called alpha cut set of the fuzzy set $A(x)$. The value of alpha cut set is x when the membership value corresponding to x is greater than or equal to the specified α value. The α-cut of a fuzzy set is denoted by $\mu_A(x)$ and can be expressed as follows:

$$A_\alpha = \{x | \mu_A(x) \geq \alpha\} \tag{4.17}$$

In Eq. (4.17) if the inequality is strict, that is if $\mu_A(x) > \alpha$, then it is called strong α-cut. The following two examples explain α-cut fuzzy set with Gaussian and triangular membership functions.

Example 4.8 Fig. 4.7 illustrates the 0.7-cut of the fuzzy set following Gaussian distribution with mean $m = 50$ and standard deviation $= 16$. Eq. (4.6) of Gaussian distribution can be expressed as

$$\mu_A(x) = e^{-\frac{1}{2}\left(\frac{x-50}{16}\right)^2}$$

The 0.7-cut can be obtained as

$$A_{0.7} = \left\{ x \big| e^{-\frac{1}{2}\left(\frac{x-50}{16}\right)^2} \geq 0.7 \right\}$$

Now,

$$e^{-\frac{1}{2}\left(\frac{x-50}{16}\right)^2} \geq 0.7$$

$$-\frac{1}{2}\left(\frac{x-50}{16}\right)^2 \geq \log(0.7)$$

$$\left(\frac{x-50}{16}\right)^2 \leq -2\log(0.7)$$

$$50 - 16\sqrt{-2\log(0.7)} \leq x \leq 50 + 16\sqrt{-2\log(0.7)}$$

$$41.0945 \leq x \leq 58.9055$$

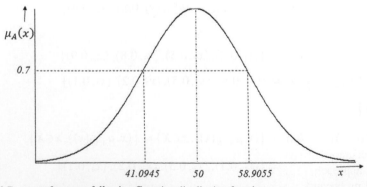

FIG. 4.7 α-cut fuzzy set following Gaussian distribution function.

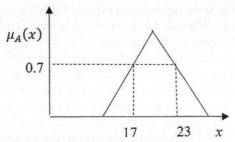

FIG. 4.8 α-cut fuzzy set following triangular distribution function.

Thus $A_{0.7} = [41.0945, 58.9055]$

The range of x is found as (41.0945, 58.9055).

Example 4.9 The previous Example 4.8 is solved with the triangular membership function given as follows.

$$\mu_A(x) = \frac{x - 10}{20 - 10}; 10 \leq x \leq 20$$

$$= \frac{30 - x}{30 - 20}; 20 \leq x \leq 30$$

The 0.7-cut can be obtained as $A_{0.7} = \{x | \mu_{0.7}(x) \geq 0.7\}$

So, $\frac{x - 10}{20 - 10} \geq 0.7$ when $10 \leq x \leq 20$.

$\frac{x}{10} - 1 \geq 0.7$ implies $x \geq 17$.

In this case, the range of x is $17 \leq x \leq 20$.

Similarly,

$$\mu_A(x) = \frac{30 - x}{30 - 20}; \quad 20 \leq x \leq 30$$

$3 - \frac{x}{10} \geq 0.7$ when $20 \leq x \leq 30$.

In this case the range of x is $20 \leq x \leq 23$

Fig. 4.8 illustrates the 0.7-cut of the triangular membership function of x combining two ranges as $A_{0.7} = [17, 23]$.

4.6 Fuzzy logic controller

Fuzzy logic controllers can be classified into two methods as Mamdani approach (Mamdani and Assilian, 1975) and the Takagi-Sugeno-Kang approach (Sugeno and Takagi, 1985). Mamdani's approach is known for its high interpretability and low accuracy. It is characterized as linguistic fuzzy modeling. Takagi-Sugeno-Kang approach (TSK approach) approach is characterized as precise fuzzy modeling because it aims to obtain high accuracy at the cost of interpretability.

Interpretability refers to the capability to express the behavior of a system in an understandable form. Whereas the accuracy of a fuzzy model indicates how closely a modeled system can be expressed.

In the case of the Mamdani Fuzzy Inference System (FIS), the consequent membership functions are also fuzzy. On the contrary, the consequent membership functions in a TSK FIS are not fuzzy (either linear or constant). Mamdani FIS more popularly used because it provides reasonably good results with a relatively simple structure and the interpretable nature of the rule base. The interpretability of TSK FIS is lacking. However, the consequent membership functions for TSK FIS can have as many parameters per rule as input variables, this translates into more degrees of freedom in the design as compared to Mamdani FIS, thus providing more flexibility in the design of the system. Mamdani FIS can be used directly for MISO systems (multiple input single output) and MIMO systems (multiple input multiple output), while the TSK FIS can only be used in the case of the MISO system.

4.6.1 Mamdani FIS

This approach was proposed by Mamdani and Assilian (1975). It is the most commonly used fuzzy approach and popularly known as the Mamdani approach. The Mamdani approach provides a systematic way to design fuzzy controllers based on linguistic rules. The Mamdani approach is a flexible and intuitive framework for designing fuzzy controllers. It allows for the incorporation of expert knowledge and linguistic descriptions into the control system, making it suitable for systems with complex and imprecise dynamics. The schematic view of the Mamdani approach fuzzy controller is illustrated in Fig. 4.9.

4.6.1.1 Fuzzification

Fuzzification is the process of transforming the crisp set into a fuzzy set. In this operation, crisp values are transformed into linguistic variables. Although, the majority of the crisp quantities are accurate, and deterministic in nature, in the real world they are liable to uncertainty or vagueness. Therefore, the vagueness can be handled by transforming the crisp quantities into linguistic variables with the help of the membership function. For example, if someone is told that the

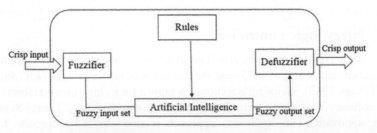

FIG. 4.9 Schematic diagram of Mamdani Fuzzy Inference System.

"outside temperature is 40°C," the person will translate the crisp value into linguistic terms through a knowledge base, and a decision for the most suitable dress for the weather will be taken.

4.6.1.2 Fuzzy rule

A fuzzy rule base is an essential component of a fuzzy inference system (FIS). It consists of a collection of fuzzy rules that define the relationship between the inputs and outputs of the system. Each rule expresses a fuzzy relationship based on the linguistic terms associated with the variables involved. The fuzzy rule base consists of a set of IF-THEN rules that relate the input variables to the output variables. Each rule specifies a condition and an associated action. The conditions are defined using fuzzy sets and membership functions, while the actions are linguistic terms that represent the desired control action. Each fuzzy rule follows an IF-THEN structure, where the IF part is called the antecedent and the THEN part is called the consequent. The antecedent specifies the conditions or criteria based on the input variables, while the consequent determines the actions or conclusions based on the output variables. The antecedent of a fuzzy rule often includes logical operators to express the relationship between multiple linguistic terms or variables. Common logical operators used in fuzzy rule bases are AND, OR, and NOT. Theoretically, there may be r^n number of rules, where n is the number of input variables having r linguistic levels. However, it is not mandatory to construct all these rules. An efficient fuzzy system can be built up with a lesser number of rules as well. Moreover, a large number of rules invites more complexity to the system. The output of each rule is also a fuzzy set. All rules are evaluated in parallel and output fuzzy sets are then aggregated into a single fuzzy set. For example, consider a fuzzy system of two inputs and a single output. In this example, only two fuzzy control rules are fired to maintain simplicity as shown in Fig. 4.10.

Rule 1: IF I_1 is A_1 and I_2 is B_1 THEN O is C_1
Rule 2: IF I_1 is A_2 and I_2 is B_2 THEN O is C_2

Fig. 4.10 illustrates the Mamdani FIS with two input and one output model. In this figure, μ_{A1} and μ_{B1} are considered as membership function values of A and B, respectively. I_1* and I_2* are the fuzzy input variables of I_1 and I_2, respectively. For rule 1, $\mu_{A1}(I_1*)$ and $\mu_{B1}(I_2*)$ are represented as the membership grade values of I_1* in A_1 and I_2* in B_1, respectively. Similarly, for rule 2, the grade of membership of I_1* in A_2 and I_2* in B_2 are represented by $\mu_{A2}(I_1*)$ and $\mu_{B2}(I_2*)$, respectively.

4.6.1.3 Fuzzy inference

The fuzzy inference process involves combining the firing strengths of the rules to determine the overall control action. There are different methods for fuzzy inference, but the most common one used in the Mamdani approach is the

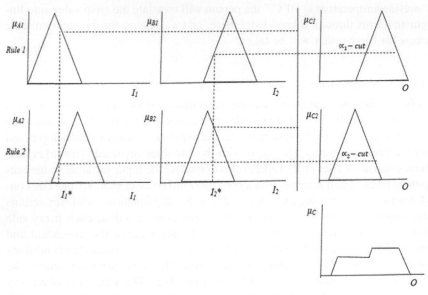

FIG. 4.10 Schematic view of working principle of a fuzzy logic controller.

max-min composition method. It computes the minimum of the firing strengths of the rules. The firing strengths of the rule 1 and 2 are calculated as follows:

$$\propto_1 = \min(\mu_{A1}(I_1^*), \mu_{B1}(I_2^*))$$
$$\propto_2 = \min(\mu_{A2}(I_1^*), \mu_{B2}(I_2^*))$$

When the inputs to the fuzzy controller are given, the fuzzy rule base is used to evaluate the firing strength of each rule. The firing strength represents the degree to which each rule is applicable to the given input values. This is determined by evaluating the degree of membership of the input variables in the fuzzy sets defined in the antecedent of each rule. The membership value of the combined control action C is shown here. This step is known as "aggregation," which is calculated using the "max" operator.

$$\mu_C(O) = \max(\mu_{c1}(O), \mu_{c2}(O))$$

4.6.1.4 Defuzzification

After the fuzzy inference step, a crisp output value needs to be obtained. Defuzzification is the process of mapping the fuzzy output sets into a crisp value that can be used to control the system. There are several methods of defuzzification, such as the centroid method, center of sums method, weighted average method, and mean of maxima method.

Centroid method: The most common method for defuzzification in the Mamdani approach is the centroid method, which calculates the center of gravity of the fuzzy output set. The calculation of defuzzified output (x^*) in the centroid method is shown as follows:

$$x^* = \frac{\sum_{i=1}^{n} A_i \times \bar{x}_i}{\sum_{i=1}^{n} A_i} \tag{4.18}$$

where A_i and $\bar{x}_i$ represents the area and centroid of ith subareas.

Fig. 4.11 shows a fuzzified output where the total area of the membership function distribution is divided into six subareas such as a, b, c, d, e, and f. The total area and centroid of the subareas are calculated and then the defuzzified value is measured using Eq. (4.18). Table 4.1 demonstrates the calculations of the area and centroid of each subareas.

The defuzzified output using the centroid method is calculated as

$$x^* = \frac{\sum_{i=1}^{n} A_i \times \bar{x}_i}{\sum_{i=1}^{n} A_i}$$

$$= \frac{1502.475}{30} = 50.08$$

Centre of sums method: The center of sums method involves the algebraic sum of individual fuzzy sets, say $\widetilde{C_1}$ and $\widetilde{C_2}$ instead of a union. The intersecting areas of the membership functions are added twice in this method. This method is similar to the weighted average method but the basic difference is that in the weighted average method, the weights are individual membership values

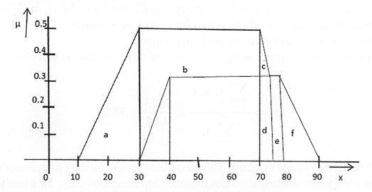

FIG. 4.11 Subareas of the fuzzified output in the centroid method.

TABLE 4.1 Area and centroid of subareas.

Number of subarea (i)	Area (A_i)	Centroid ($\bar{x}_i$)	$A_i \times \bar{x}_i$
a	$\frac{1}{2} \times 20 \times 0.5 = 5$	$\frac{10 + 30 + 30}{3} = 23.33$	116.65
b	$(70 - 30) \times 0.5 = 20$	$\frac{70 + 30}{2} = 50.00$	1000.00
c	$\frac{1}{2} \times (75 - 70) \times 0.2 = 0.50$	$\frac{70 + 70 + 75}{3} = 71.66$	35.83
d	$5 \times 0.3 = 1.50$	$\frac{70 + 75}{2} = 72.50$	108.75
e	$5 \times 0.3 = 1.50$	$\frac{75 + 80}{2} = 77.50$	116.25
f	$\frac{1}{2} \times 10 \times 0.3 = 1.50$	$\frac{80 + 80 + 90}{3} = 83.33$	124.995
Total	30	378.32	1502.475

whereas in the center of sums method, the weights are areas of the respective membership functions. Suppose n is the total number of fired rules for a given input. The firing area of ith fired rule is denoted by A_i and $\bar{x}_i$ is the centre of the area A_i. Then according to this method, the crisp value of the given input is

$$x^* = \frac{\sum\limits_{i=1}^{n} A_i \times \bar{x}_i}{\sum\limits_{i=1}^{n} A_i} \qquad (4.19)$$

The aggregated fuzzy set of two fuzzy sets $\widetilde{C_1}$ and $\widetilde{C_2}$ is shown in Fig. 4.12, which illustrates the area of these two fuzzy sets as A_1 and A_2. Table 4.2 demonstrates the area and center of area calculations.

The defuzzified output using the center of sums method is calculated as

$$x^* = \frac{\sum\limits_{i=1}^{n} A_i \times \bar{x}_i}{\sum\limits_{i=1}^{n} A_i} = \frac{2275}{42.5} = 53.53$$

Weighted average method: The weighted average method of defuzzication can be used for symmetrical output membership functions only. This method is computationally less expensive. The largest value of each membership function

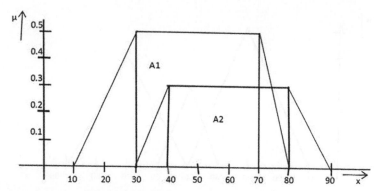

FIG. 4.12 Areas of the fuzzified output in the center of sums method.

TABLE 4.2 Area and center of area of fuzzy sets.

Fuzzy sets	Area (A_i)	Centre of area ($\bar{x}_i$)	$A_i \times \bar{x}_i$
$\tilde{C}_1$	$A_1 = \frac{1}{2} \times (80 - 10) + (70 - 30) \times 0.5 = 27.5$	$x_1 = \dfrac{70 + 30}{2} = 50$	1375
$\tilde{C}_2$	$A_2 = \frac{1}{2} \times (90 - 30) + (80 - 40) \times 0.3 = 15$	$x_2 = \dfrac{80 + 40}{2} = 60$	900
Total	42.5	110	2275

is obtained in this method. The defuzzified output is measured using the expression as follows:

$$x^* = \frac{\sum \mu(x)x}{\sum \mu(x)} \tag{4.20}$$

where x is the element with the maximum membership function.

For example, we consider the fuzzified output given in Fig. 4.13.

The defuzzified output value using the weighted average method is calculated as

$$x^* = \frac{(30 \times 0.5) + (50 \times 0.3) + (70 \times 0.4)}{(0.5 + 0.3 + 0.4)} = \frac{58}{1.2} = 48.33$$

Mean of maxima method: This method is also called the middle of maxima method. In this method, the maximum membership need not be a single point. The range of output variables corresponding to the maximum value of membership is calculated. The mean value of the maxima is considered the crisp output value.

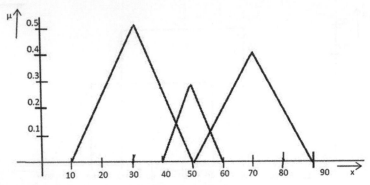

FIG. 4.13 Fuzzified output in weighted average method.

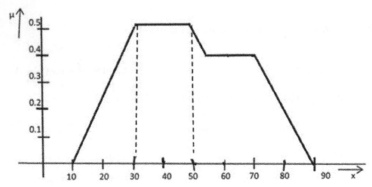

FIG. 4.14 Aggregated fuzzified output in the mean of maxima method.

From Fig. 4.14, the defuzzified value of the mean of maxima is calculated as

$$x^* = \frac{30 + 50}{2} = 40$$

4.6.1.5 Fuzzy inference system with step-by-step example using MATLAB

Two inputs and one output fuzzy system are considered here for a step-by-step explanation of Mamdani's fuzzy inference method using MATLAB Fuzzy Logic Toolbox. Suppose, x_1 and x_2 are two input variables and y is considered as output variable in crisp form. The "Fuzzy" command in the MATLAB command line is used to open Fuzzy Inference System. Fig. 4.15 demonstrates the FIS editor of MATLAB Fuzzy Logic Toolbox used to execute this example. The FIS editor is capable of handling the number of input and output variables. The fuzzification of the inputs and output is performed using the Membership Function Editor as illustrated in Figs. 4.16–4.18. Each input is represented by three linguistic terms such as low, medium, and high, and the output y is fuzzified into five linguistic

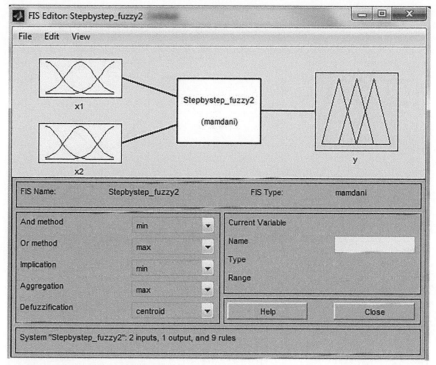

FIG. 4.15 FIS editor of MATLAB.

levels such as very good, good, average, poor, and very poor. The name, range, and type of the membership functions are accredited using the Membership Function Editor. The triangular membership function is used in this example for fuzzification of input and output variables. As there are two input variables and each one of them has three linguistic levels, hence the total of 9 ($=3^2$) number of rules formed. Table 4.3 is formed based on the expert's knowledge base that is used to construct 9 rules. The Rule Editor of the MATLAB Fuzzy Toolbox is used to incorporate rules in the FIS as shown in Fig. 4.19.

Rule 1: IF x_1 is high AND IF x_2 is high THEN y is very good
Rule 2: IF x_1 is high AND IF x_2 is medium THEN y is very good
Rule 3: IF x_1 is high AND IF x_2 is low THEN y is average
Rule 4: IF x_1 is medium AND IF x_2 is high THEN y is good
Rule 5: IF x_1 is medium AND IF x_2 is medium THEN y is good
Rule 6: IF x_1 is medium AND IF x_2 is low THEN y is average
Rule 7: IF x_1 is low AND IF x_2 is high THEN y is average
Rule 8: IF x_1 is low AND IF x_2 is medium THEN y is poor
Rule 9: IF x_1 is low AND IF x_2 is low THEN y is very poor

Let fuzzy controller processing for the input variables x_1^* and x_2^* be 16 and 180, respectively. Fig. 4.20 refers to the first input (x_1^*) belonging to the "high"

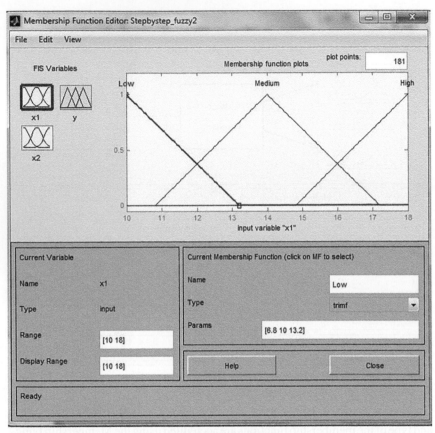

FIG. 4.16 Input variable *x1* in Membership Function Editor.

and "medium" linguistic levels with the same membership value, i.e., 0.375. The second input (x_2^*) has higher membership for the "medium" linguistic level and low belongingness for the "high" membership function as illustrated in Fig. 4.21. Similarly, the second input 180 belongs to the "high" level with 0.286 membership and the "medium" level with 0.536 membership value, respectively. Using the principle of a similar triangle, the membership functions of x_1^* and x_2^* are calculated as follows:

$$\frac{x}{1} = \frac{17.2 - 16}{17.2 - 14} \rightarrow \mu_{\text{medium}}(x_1^*) = 0.375$$

$$\frac{x}{1} = \frac{16 - 14.8}{18 - 14.8} \rightarrow \mu_{\text{high}}(x_1^*) = 0.375$$

$$\frac{x}{1} = \frac{180 - 172}{200 - 172} \rightarrow \mu_{\text{high}}(x_2^*) = 0.286$$

$$\frac{x}{1} = \frac{180 - 165}{193 - 165} \rightarrow \mu_{\text{medium}}(x_2^*) = 0.536$$

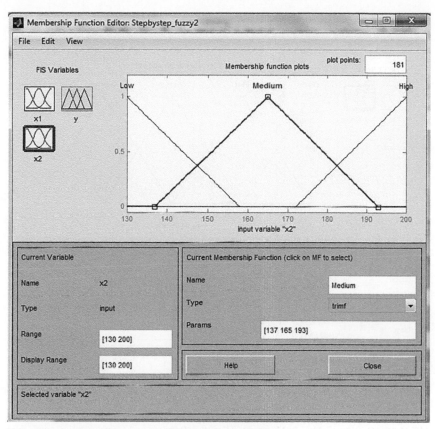

FIG. 4.17 Input variable x_2 in Membership Function Editor.

The fired rules and output of the FIS can be seen from the Rule viewer of the Fuzzy Logic Toolbox. In this example, only 4 rules such as rule number 1, 2, 4, and 5 are fired out of 9 rules as shown in Fig. 4.22. The "min" operator is used on the antecedent to obtain the output function.

The firing strengths of the rules ($\propto$) are calculated as follows:

$$\propto_1 = \min\left(\mu_{\text{high}}\left(x_1{}^*\right), \mu_{\text{high}}\left(x_2{}^*\right)\right) = \min(0.375, 0.286) = 0.286$$

$$\propto_2 = \min\left(\mu_{\text{high}}\left(x_1{}^*\right), \mu_{\text{medium}}\left(x_2{}^*\right)\right) = \min(0.375, 0.536) = 0.375$$

$$\propto_3 = \min\left(\mu_{\text{medium}}\left(x_1{}^*\right), \mu_{\text{high}}\left(x_2{}^*\right)\right) = \min(0.375, 0.286) = 0.286$$

$$\propto_4 = \min\left(\mu_{\text{medium}}\left(x_1{}^*\right), \mu_{\text{medium}}\left(x_2{}^*\right)\right) = \min(0.375, 0.536) = 0.375$$

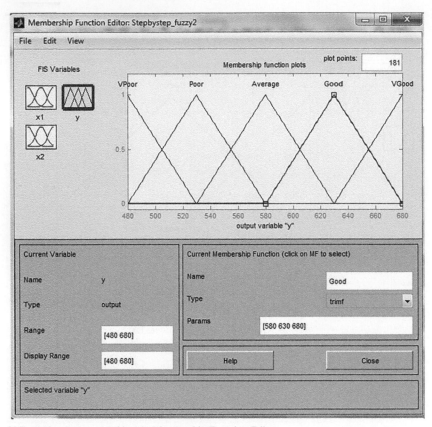

FIG. 4.18 Output variable y in Membership Function Editor.

TABLE 4.3 Rule knowledge base.

		X_1		
		High	*Medium*	*Low*
X_2	High	Very good	Good	Average
	Medium	Very good	Good	Poor
	Low	Average	Average	Very poor

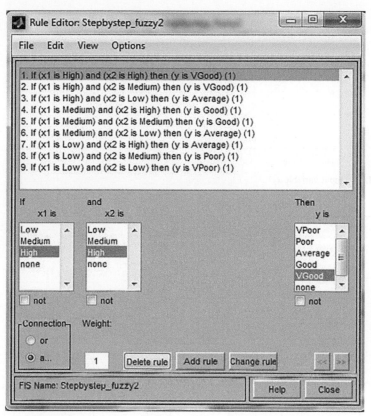

FIG. 4.19 Rule editor.

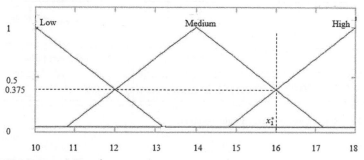

FIG. 4.20 Input variable $x_1{}^*$.

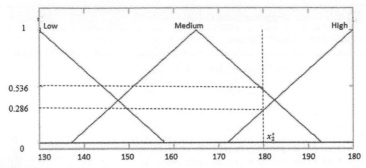

FIG. 4.21 Input variable $x_2{}^*$.

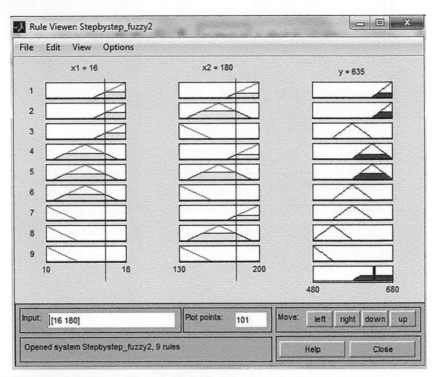

FIG. 4.22 Rule viewer of Fuzzy Logic Toolbox.

The "max" operator is used to compute the aggregated fuzzy output. The aggregation of the fired output functions is shown in Fig. 4.22. The crisp output is then obtained using three different defuzzification methods: center of sums method, centroid method, and mean of maxima method. The weighted average method is not applicable for this example as the output fuzzy set is not symmetrical.

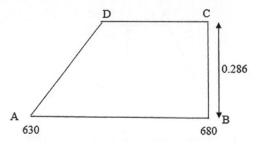

FIG. 4.23   Fuzzified output of 1st fired rule.

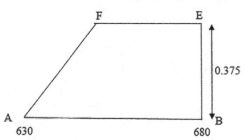

FIG. 4.24   Fuzzified output of 2nd fired rule.

Centre of sums method: For the 1st fired rule, the output fuzzy set is given by the trapezium ABCD as shown in Fig. 4.23. The height BC represents the firing strength ($\propto_1$) that is 0.286. The distance between CD is calculated as follows:

$$\frac{CD}{680 - 630} = (1 - 0.286)$$
$$CD = 35.7$$

For the 2nd fired rule, the output fuzzy set is given by the trapezium ABEF as shown in Fig. 4.24, the height BE representing the firing strength ($\propto_2$) that is 0.375. The length EF is calculated from the geometry of a similar triangle shown in this figure.

$$\frac{EF}{680 - 630} = (1 - 0.375)$$
$$EF = 31.25$$

For the 3rd and 4th fired rules, the regions are given by the trapeziums MNOP and MNQR, respectively. The firing strengths ($\propto_3$ and $\propto_4$) are measured as 0.286 and 0.375. The length OP and QR are calculated from the geometry of a similar triangle as shown in Figs. 4.25 and 4.26, respectively

$$\frac{OP}{680 - 580} = (1 - 0.286)$$
$$OP = 71.4$$

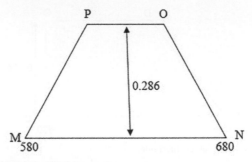

FIG. 4.25 Fuzzified output of 3rd fired rule.

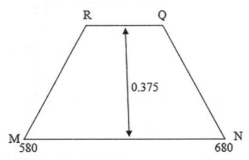

FIG. 4.26 Fuzzified output of 4th fired rule.

And

$$\frac{QR}{680 - 580} = (1 - 0.375)$$
$$QR = 62.5$$

The mid-point of MN, i.e., 630 is the centre of the area for both the output fuzzy sets MNOP and MNQR. Table 4.4 shows the areas and centers of the areas of the output fuzzy sets ABCD, ABEF, MNOP, and MNQR.

Thus, the defuzzified crisp value in the center of sums method is calculated as

$$\frac{658.376 \times 12.2551 + 659.327 \times 15.2343 + 630 \times (24.5102 + 30.4687)}{12.2551 + 15.2343 + 24.5102 + 30.4687}$$
$$= 639.6343$$

Centroid method: For this given input, the fuzzified output corresponds to the region R1 and region R2 shown in Fig. 4.27. Suppose the value of the point B in this figure is x. From a similar triangle, x is calculated as

$$\frac{x - 580}{0.375} = \frac{630 - 580}{1}$$
$$\therefore x = 598.75$$

The center and area of the regions are provided in Table 4.5.

TABLE 4.4 Area and center of the area of the fuzzy output.

Regions	Area (A_i)	Centre of the area ($\bar{x}_i$)
ABCD	$\frac{1}{2}(35.7 + (680 - 630)) \times 0.286 = 12.2551$	$680 - \dfrac{(35.7)^2 + (680 - 630)^2 + (35.7 \times (680 - 630))}{3 \times (35.7 + (680 - 630))} = 658.376$
ABEF	$\frac{1}{2}(31.25 + (680 - 630)) \times 0.375 = 15.2343$	$680 - \dfrac{(31.25)^2 + (680 - 630)^2 + (31.25 \times (680 - 630))}{3 \times (31.25 + (680 - 630))} = 659.327$
MNOP	$\frac{1}{2}(71.4 + (680 - 580)) \times 0.286 = 24.5102$	630
MNQR	$\frac{1}{2}(62.5 + (680 - 580)) \times 0.375 = 30.4687$	630

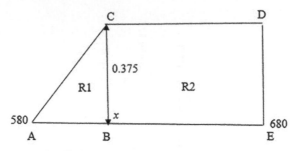

FIG. 4.27 Aggregated fuzzified output.

TABLE 4.5 Area and center of the area of the fuzzy output.

Marked regions	Area (A_i)	Centre of area $(\bar{x}_i)$
R1	$\frac{1}{2}(598.75 - 580) \times 0.375 = 3.5156$	$\frac{(598.75 \times 2) + 580}{3} = 592.5$
R2	$(680 - 598.75) \times 0.375 = 30.4687$	$\frac{598.75 + 680}{2} = 639.375$

Thus, the defuzzified crisp value using the centroid method comes out as

$$\frac{3.5156 \times 592.5 + 30.4686 \times 639.375}{(3.5156 + 30.4687)} = 634.5258$$

Mean of maxima method: The fuzzified output of the four fired rules is already provided in ACDE as shown in Fig. 4.27. In this figure, the maximum value of the firing strength is 0.375, which occurs in a range of 598.75 to 680. The mean of 598.75 and 680 is found to be 639.375. Therefore, the defuzzified crisp value following the mean of maxima method is 639.375.

4.6.1.6 Application of fuzzy logic controller on silk quality prediction

Sericulture is an agro-based industry. It involves the rearing of silkworms for the production of raw silk, which is obtained out of silk cocoons spun by certain species of insects. Unlike other natural fibers, sericulture is a unique combination of botanical, zoological, and textile sciences dealing with on and off-farm activities. Hence, the handling of silk textiles is a little delicate (Kar et al., 2018). It is well recognized as a soft science where there is always some amount of uncertainty or vagueness or fuzziness that prevails in the underlying phenomenon. It has been experienced that in most of the cases, silk cocoon characteristics are not properly reflected in the raw silk quality and it cannot be justified with simple statistical analysis because of the presence of a lot of ambiguities or lack of clarity in the objective values. Since cocoon is a biomaterial, the inherent relationship between cocoons and silk yarn quality is absolutely imprecise. The quality of cocoons plays a very important role in reeling and subsequently has a great bearing on the end product as well.

In this study total of 25 lots of mulberry silk cocoons of the multi-bivoltine category were collected from the cocoon market. The defective cocoon percentage (DC%), shell ratio (SR), cocoon weight, i.e., the weight of 100 cocoons in gram, and cocoon volume, i.e., number of cocoons per liter were measured using the standard technique. The SR (%) was estimated as the ratio of the shell weight to the cocoon weight expressed as a percentage. The hundred cocoons were randomly selected from each lot and the average SR (%) was calculated. The DC (%) was measured as a percentage of defective cocoons by a number from a lot of 1 kg of cocoons. Cocoon volume was determined by number of cocoons that can be accommodated in a container of 1 L. The lower the number of cocoons per liter, the higher is the average cocoon size. The cocoon quality improves with the rise of SR and cocoon weight values whereas it becomes inferior with the upsurge of DC (%) and cocoon volume. The experimentally tested DC%, SR%, cocoon weight, and cocoon volume were used as input parameters, and filament length (FL) was used as output parameter. The experimentally collected data from 25 lots of silk cocoons are tabulated in Table 4.6.

The Mamdani fuzzy logic controller was used to predict the quality of silk cocoons in terms of filament length. Three linguistic fuzzy sets namely low, medium, and high are chosen for each of the input parameters in such a way that they are equally spaced and cover the whole input spaces. The triangular membership functions were used for the fuzzification of inputs and output parameters. The selection of the triangular membership function is attributable to its simplicity as it is merely a collection of three points forming a triangle. Figs. 4.28–4.31 depict the triangular membership curves for DC, SR, cocoon volume, and cocoon weight, respectively. The fuzzification of the output parameter, i.e., FL was attributed using five linguistic levels such as very poor, poor, average, good, and very good. Fig. 4.32 shows the triangular membership curve for FL. Theoretically, there could be $3^4 = 81$ fuzzy rules, as there were four input variables and each one of them had three linguistic levels. However, to simplify the expert system only 40 fuzzy rules were manually constructed using the expert's knowledge as shown in Table 4.7. The fuzzy rules were formed using IF-THEN statements and the "AND" operator was used to connect the input variables. Rule 1 is shown here as an example:

IF DC is low AND IF SR is low AND IF cocoon weight is high AND IF cocoon volume is high THEN filament length is average

TABLE 4.6 Silk cocoons data.

Lot no.	D.C. (%)	S.R. (%)	Cocoon volume (no./Lt)	Cocoon weight (g/100 cocoons)	Filament length (m)
1	9.3	15.32	143	141	577
2	7.4	14.7	142	138	601
3	9.8	14.87	140	143	599
4	8.9	14.96	138	142	636
5	11.13	17.66	131	195	605
6	10.53	17.63	137	181	605
7	11.78	17.86	141	178	589
8	10.47	18.01	135	173	598
9	10.63	17.22	142	182	602
10	11.69	17.05	136	179	591
11	11.75	17.62	139	171	596
12	10.36	17.68	139	168	589
13	10.97	17.83	141	172	592
14	11.12	17.43	144	165	585
15	11.14	16.23	128	183	637
16	12.44	17.34	136	191	616
17	12.43	16.47	145	179	628
18	9.84	15.62	132	167	599
19	9.89	15.94	141	178	577
20	8.52	16.04	155	145	663
21	10.97	16.12	127	167	648
22	10.82	15.37	142	176	621
23	7.47	16.73	136	154	647
24	8.19	16.41	138	159	653
25	8.21	18.01	148	153	657

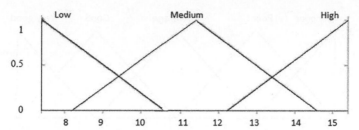

FIG. 4.28 Triangular membership functions of DC%.

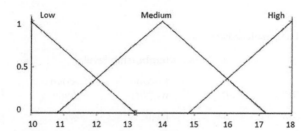

FIG. 4.29 Triangular membership functions of SR%.

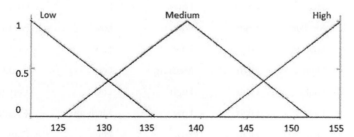

FIG. 4.30 Triangular membership functions of cocoon volume.

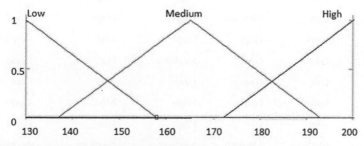

FIG. 4.31 Triangular membership functions of cocoon weight.

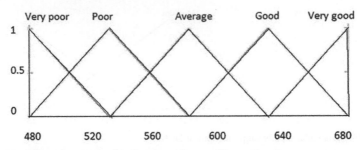

FIG. 4.32 Triangular membership functions of output filament length.

TABLE 4.7 Fuzzy rule base.

Rule no.	Membership level				
	DC%	SR	Cocoon weight	Cocoon volume	FL
1	Low	Low	High	High	Average
2	High	High	Low	Low	Good
3	High	Medium	Low	Low	Average
4	High	Low	Low	Low	Poor
5	Medium	High	Low	Low	Average
6	Low	High	Low	Low	Good
7	High	High	Medium	Low	Good
8	High	High	High	Low	Very good
9	High	High	Low	Medium	Average
10	High	High	Low	High	Poor
11	Medium	Low	High	High	Average
12	High	Low	High	High	Poor
13	Low	Medium	High	High	Good
14	Low	High	High	High	Very good
15	Low	Low	Medium	High	Good
16	Low	Low	Low	High	Poor
17	Low	Low	High	Medium	Good
18	Low	Low	High	Low	Good
19	Medium	Medium	Medium	Medium	Average
20	Low	Medium	Medium	Medium	Good

TABLE 4.7 Fuzzy rule base—cont'd

Rule no.	Membership level				
	DC%	SR	Cocoon weight	Cocoon volume	FL
21	High	Medium	Medium	Medium	Poor
22	Medium	Low	Medium	Medium	Average
23	Medium	High	Medium	Medium	Good
24	Medium	Medium	Low	Medium	Average
25	Medium	Medium	High	Medium	Good
26	Medium	Medium	Medium	Low	Good
27	Medium	Medium	Medium	High	Poor
28	High	High	Medium	Medium	Good
29	High	Medium	Low	Medium	Poor
30	High	Medium	Medium	Low	Average
31	Medium	High	High	Low	Very good
32	Low	Low	Medium	Medium	Average
33	Low	Medium	High	Medium	Good
34	Medium	High	Low	Medium	Average
35	Medium	High	Medium	Low	Good
36	Medium	Medium	Low	Low	Average
37	Low	High	High	Medium	Very good
38	Low	High	High	Low	Very good
39	High	Medium	Low	High	Poor
40	Medium	High	High	High	Good

The fuzzy logic controller toolbox of MATLAB 7.11 is used to solve this problem as shown in Fig. 4.33. The fuzzy logic controller simultaneously acted upon the whole set of predefined 40 rules for four fuzzified input parameters such as 9.3, 15.3, 143, and 141 as DC%, SR%, weight, and volume, respectively. The rules were fired based on the linguistic levels of input parameters shown in Fig. 4.34. The fuzzy sets that represent the outputs of each rule were combined into a single fuzzy set by the process of aggregation. The input of the aggregation process was the list of truncated output functions evaluated by each rule. The "max" function was used to aggregate the output of each rule into a

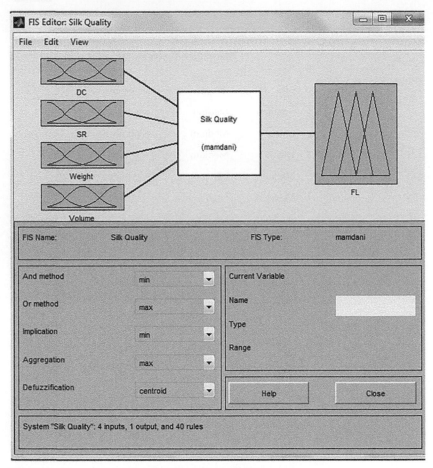

FIG. 4.33 Fuzzy logic toolbox for silk quality prediction.

single fuzzy set of the output variable. The aggregate fuzzified output was then converted into a single crisp value by the process of defuzzification. The centroid method that returns the center of the area under the curve was used as a defuzzification method to yield the crisp output of FL. The filament length obtained from this fuzzy model is 616 m.

For validation of the model, the input datasets were used to predict the filament length of the silk cocoon using the developed fuzzy logic controller. The obtained FL predicted by the fuzzy logic controller was compared with the actual FL dataset. The validation datasets are represented in Table 4.8, which infers that the mean error percentage between actual and predicted filament length is 4.76%. The low mean error percentage of FL elucidates a reasonably good degree of prediction consistency of the fuzzy logic controller.

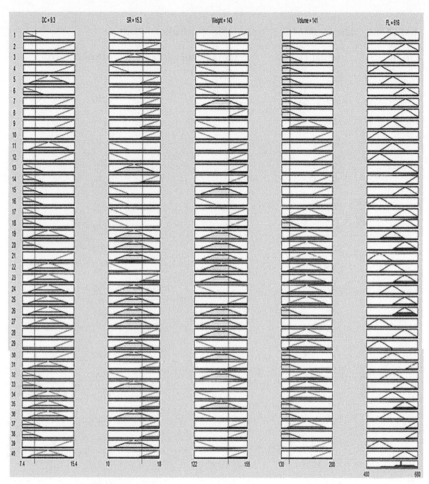

FIG. 4.34 Fuzzy rules and output.

TABLE 4.8 Validation of the fuzzy logic controller.

Lot no.	Actual FL (m)	Obtained FL (m)	Absolute error (%)
1	577	616	6.76
2	601	630	4.83
3	599	615	2.67
4	636	608	4.40
5	605	580	4.13

Continued

TABLE 4.8 Validation of the fuzzy logic controller—cont'd

Lot no.	Actual FL (m)	Obtained FL (m)	Absolute error (%)
6	605	630	4.13
7	589	630	6.96
8	598	629	5.18
9	602	629	4.49
10	591	619	4.74
11	596	630	5.70
12	589	630	6.96
13	592	630	6.42
14	585	630	7.69
15	637	580	8.95
16	616	630	2.27
17	628	594	5.41
18	599	598	0.17
19	577	587	1.73
20	663	637	3.92
21	648	592	8.64
22	621	582	6.28
23	647	630	2.63
24	653	630	3.52
25	657	660	0.46
Mean error %			4.76

4.6.2 Takagi-Sugeno-Kang FIS

This fuzzy inference system was introduced by Takagi, Sugeno, and Kang (TSK) in 1985. Popularly known as the Sugeno fuzzy inference system. This method is similar to the Mamdani Fuzzy Inference System in many aspects. The fuzzification and application to fuzzy operators are similar to Mamdani FIS. The basic difference between this method and with Mamdani approach lies in the consequent of fuzzy rules. TSK method employs the linear function of input variables as rule consequent whereas the Mamdani approach uses the fuzzy sets as rule consequent. The output membership function in TSK FIS is either linear or constant. The basic format of the rule in this model is shown as follows:

$$\text{IF } x \text{ is } A \text{ AND } y \text{ is } B \text{ THEN } O = f\,(x, y),$$

where A and B are fuzzy sets in the antecedent; $O = f(x, y)$ is a crisp function in the consequent.

Generally, $f(x, y)$ is a polynomial in the input variables x and y. If $f(x, y)$ is a first-order polynomial, then it is a first-order fuzzy model. TSK method becomes a zero-order fuzzy model, if f is a constant.

Fig. 4.35 illustrates a Sugeno fuzzy model where x and y are the values of input 1 and input 2, respectively. Each rule generates z and w as output and rule strength values, respectively.

The rule output is $z = ax + by + c$

where a, b, and c are constant coefficients. For the zero-order TSK fuzzy model output become constant as $a = b = 0$.

The firing strength (w) of the rule is derived from the rule antecedent as

$$w = \text{AndMethod}(F_1(x)F_1(y)) \qquad (4.21)$$

where $F_1(x)$ and $F_1(y)$ are the membership functions of input 1 and input 2, respectively.

The output of each rule is a product of z_i and w_i. The final output of the system is the weighted average of the overall rule output.

$$\text{Final Output} = \frac{\sum_{i=1}^{n} w_i z_i}{\sum_{i=1}^{n} w_i} \qquad (4.22)$$

where, i and n represent rule number and total number of rules, respectively.

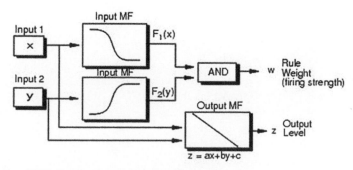

FIG. 4.35 Takagi-Sugeno-Kang fuzzy inference system.

4.7 Summary

Fuzzy logic, a mathematical framework that deals with uncertainty and imprecision, proves to be highly beneficial in handling the complex and nuanced nature of textile manufacturing. The chapter begins by introducing the fundamentals of fuzzy logic, highlighting its ability to handle vague and subjective information. It defines fuzzy set mathematics and compares it with the classical set. The working principle of fuzzy logic is methodically explicated with step-by-step examples. The chapter discusses about implementation of the fuzzy inference system, which includes linguistic variables, membership functions, and rule bases. It also explains the fuzzy logic controller with the MATLAB toolbox. It then delves into the specific application of fuzzy logic in the textile domain.

References

Kar, N. B., Das, S., Ghosh, A., & Banerjee, D. (2018). Silk cocoons quality characterization by fuzzy control chart. *Industrial Engineering Journal*, *11*(12), 1–11.

Mamdani, E. H., & Assilian, S. (1975). An experiment in linguistic synthesis with a fuzzy logic controller. *International Journal of Man-Machine Studies*, *7*(1), 1–13.

Sivanandam, S. N., Sumathi, S., & Deepa, S. N. (2010). *Introduction to fuzzy logic using MATLAB*. Berlin, Heidelberg: Springer.

Sugeno, M., & Takagi, T. (1985). Fuzzy identification systems and its application to modelling and control. *IEEE Trans on Systems, Man and Cybernetics*, *15*, 116–132.

Zadeh, L. A. (1965). Fuzzy sets. *Information and Control*, *8*(3), 338–353.

Zimmerman, H. J. (1996). *Fuzzy set theory and its applications* (3rd ed., pp. 1–456). Boston: Kluwer Academic Publishers.

Chapter 5

Rough set

5.1 Introduction

The concept of rough sets is a new mathematical approach to deal with imperfect knowledge in the field of artificial intelligence. The most successful approach to handle imprecise knowledge is the fuzzy set theory proposed by Zadeh (1965). Rough set theory was introduced by the Polish mathematician and computer scientist Prof. Pawlak in the early 1980s. Rough set theory provides a powerful framework for dealing with uncertainty and vagueness in data, making it a valuable tool for knowledge discovery and decision support in various domains. It has fundamental importance to artificial intelligence and cognitive sciences, especially in the areas of machine learning, decision analysis, knowledge acquisition, knowledge discovery from datasets, expert systems, and pattern recognition (Pawlak, 1982).

At its core, rough set theory is concerned with the representation and analysis of imprecise and incomplete information. It provides a formal method for dividing a universe of objects into distinct classes or categories based on the available information. This division is done by discerning which attributes or features are essential for distinguishing between different classes of objects and which attributes are redundant or superfluous.

The central idea behind rough sets is to create approximations of sets using lower and upper approximations. Lower approximation represents the set of objects that can be confidently classified as belonging to a specific category, while upper approximation includes objects that may or may not belong to the category. The difference between these two approximations is referred to as the boundary region, where uncertainty exists.

Rough set theory has found applications in various fields, including data mining, pattern recognition, expert systems, and knowledge representation. Its ability to handle imperfect or incomplete information has made it particularly valuable in real-world problems where data is often noisy or uncertain.

In this chapter, we will delve into the fundamental concepts of rough sets, including formal definitions, mathematical representations, and basic operations. We will also explore practical applications and discuss how rough set theory can be used to extract valuable knowledge from complex and uncertain data. Whether you are a researcher, data scientist, or practitioner in the field of

Artificial Intelligence in Textile Engineering. https://doi.org/10.1016/B978-0-443-15395-2.00005-3

143

artificial intelligence, this chapter aims to provide you with a solid foundation in rough set theory and its potential for solving real-world problems.

5.2 Basics of rough set theory with step-by-step examples

Rough set theory can be observed as a specific implementation of Frege's idea of vagueness (Pawlak, 1982) where imprecision is expressed by a boundary region of a set, and not by membership functions like in fuzzy set theory. Prof. Pawlak introduced the concept of rough sets theory to represent vagueness, ambiguity, and uncertainty (Pawlak, 1982). It is defined by equivalence relation based on lower and upper approximation. Rough set theory provides a systematic approach for processing and representing vague concepts caused by indiscernibility in situations with incomplete information or a lack of knowledge. The basic concepts of rough set theory are discussed in the following section.

5.2.1 Information system

Definition 5.1 An information system is an ordered 4-tuple (U, Q, V, f) where U is the set of objects, Q is the set of features or attributes, V is the set of values of the features or attributes, i.e., $V = \cup_{q \in Q} V_q$ and f is a function from $U \times Q \rightarrow V$.

In the earlier definition V_q represents the set of all possible values of the attribute q.

5.2.2 Decision table and decision algorithm

The decision table can be classified into two classes of attributes, called condition and decision attributes. Each row of a decision table determines a decision rule, which specifies decisions that should be taken when conditions pointed out by condition attributes are satisfied. Sometimes decision rules have the same conditions but different decisions. Such rules are called inconsistent (nondeterministic, conflicting); otherwise, the rules are referred to as consistent (certain, deterministic, nonconflicting). Sometimes consistent decision rules are called sure rules, and inconsistent rules are called possible rules. Decision tables containing inconsistent decision rules are called inconsistent (nondeterministic, conflicting); otherwise, the table is consistent (deterministic, nonconflicting).

A set of decision rules is called a decision algorithm because for each decision table it can be associated with the decision algorithm, consisting of all the decision rules. The distinction between decision tables and decision algorithms is that the decision table is a collection of datasets, whereas a decision algorithm is a collection of implications, that is, logical expressions (Pawlak, 1991).

5.2.3 Indiscernibility

The philosophy of rough set is founded on the assumption that with every object of the universe of discourse some information (data, knowledge) is associated. Objects characterized by the same information are indiscernible (similar) in view of the available information about them. The indiscernibility relation is the mathematical basis of rough set theory and is considered as a relation between two objects or more where all the values are identical in relation to a subset of considered attributes. The indiscernibility relation is intended to express the fact that due to the lack of knowledge it is unable to discern some objects employing the available information. Approximation is another important concept in rough set theory, being associated with the meaning of the approximations in topological operations (Wu et al., 2004).

A set of objects in an information system may be similar in the sense that there exists a set of features $P \subseteq Q$ such that given any feature q belonging to P, the value of it for every object in the said set of objects is the same. These objects are called P-indiscernible.

Definition 5.2 Suppose (U, Q, V, f) is an information system and $P \subseteq Q$. A relation $\widetilde{P}$ with respect to P on $U \times U$ is said to be indiscernible if for any $x_i, x_j \in U$, $x_i \widetilde{P} x_j$ if and only if

$$\forall q \in P; f(x_i, q) = f(x_j, q)$$

It can be shown that the relation $\widetilde{P}$ is reflexive, symmetric, and transitive, and hence it is an equivalence relation. The equivalence relation $\widetilde{P}$ forms equivalence classes with respect to the objects in U. These equivalence classes are disjoint and the space U is the union of all such distinct equivalence classes.

Example 5.1 Table 5.1 shows the fiber and yarn dataset. Six inputs such as fiber strength (FS), fiber elongation (FE), upper half mean length (UHML), uniformity index (UI), fiber fineness (FF), and yarn fineness (YF) were considered. Yarn tensile strength (T) was considered as an output parameter. Table 5.1 represents 11 lots of cotton fiber properties along with corresponding yarn strength.

In this case, the indiscernibility with respect to the features say {FS, FE} is given by

$$\text{Ind}(\text{FS}, \text{FE}) = [\{x_1, x_4, x_7\}, \{x_2, x_8\}, \{x_{10}, x_{11}\}, \{x_5, x_6\}, \{x_3\}, \{x_9\}]$$

Similarly, $\text{Ind}(\text{FS}, \text{FE}, \text{UHML}) = [\{x_1, x_4, x_7\}, \{x_2, x_8\}, \{x_3\}, \{x_5\}, \{x_6\}, \{x_9\}, \{x_{10}, x_{11}\}]$.

5.2.4 Lower and upper approximation

The classical set is a primitive notion and is defined intuitively or axiomatically. Contrarily the fuzzy sets contain objects that satisfy imprecise properties of

TABLE 5.1 Dataset.

Lots	FS	FE	UHML	UI	FF	Ne	Tenacity (T)
x_1	28.7	6.5	1.09	81.0	4.4	10.2	14.62
x_2	28.5	6.6	1.15	80.2	3.5	10.1	14.04
x_3	30.8	6.4	1.13	82.6	4.3	10.0	14.76
x_4	28.7	6.5	1.09	81.0	4.4	29.8	12.33
x_5	27.5	6.3	1.07	80.0	4.5	10.1	13.7
x_6	27.5	6.3	1.09	80.0	4.3	21.8	13.27
x_7	28.7	6.5	1.09	81.0	4.4	29.8	12.33
x_8	28.5	6.6	1.15	80.2	3.5	21.8	12.93
x_9	29.2	5.3	0.98	80.0	4.5	29.8	12.72
x_{10}	29	6.7	1.05	81.9	4.2	10.0	14.67
x_{11}	29	6.7	1.05	81.9	4.5	21.5	13.48

membership, which is not a matter of affirmation or denial, but a matter of a degree of belongingness. Rough sets are defined by approximations that are interior and closure operations in a topology generated by indiscernibility relation. The details are furnished as follows.

Definition 5.3 For any $x \in U$, the equivalence class of x denoted by $[x]_{\widetilde{P}}$ with respect to $\widetilde{P}$ is

$$[x]_{\widetilde{P}} = \left\{ y \in U | y \widetilde{P} x \right\}$$

Now we can define the lower and upper approximation of a given subset X of U.

Definition 5.4 P-lower approximation of X: Let X be a subset of U and P be a relation on U. Then the P-lower approximation of X is

$$P_*(X) = \bigcup_{x \in U} \{P(x) | P(x) \subseteq X\}$$

where $P(x)$ is the equivalence class of x with respect to the relation P.

Example 5.2 We consider the dataset as in Example 5.1.

In Example 5.1, consider a set $X = \{x_2, x_3, x_4, x_5, x_6, x_7, x_8\}$ and the indiscernibility relation (P) with respect to the feature subset (FS, FE). Then, the P-lower approximation is given by,

$$P_*(X) = \{x_2, x_8\} \cup \{x_3\} \cup \{x_5, x_6\} = \{x_2, x_3, x_5, x_6, x_8\}$$

Definition 5.5 P-upper approximation of X: Let X be a sub set of U and P be a relation on U. Then the P-upper approximation of X is

$$P^*(X) = \bigcup_{x \in U} \{P(x) | P(x) \cap X \neq \varphi\}$$

Example 5.3 Consider the same set and same indiscernible relation as in Example 5.2.

Then the P-upper approximation of X is given by,

$$P^*(X) = \{x_1, x_4, x_7\} \cup \{x_2, x_8\} \cup \{x_5, x_6\} \cup \{x_3\} = \{x_1, x_2, x_3, x_4, x_5, x_6, x_7, x_8\}$$

Definition 5.6 P-boundary region of X: Let X and P are as given in the previous two definitions. Then the P-boundary region of X is

$$B_P(X) = P^*(X) - P_*(X)$$

Example 5.4 Here we consider the set X and the indiscernibility relation as in Examples 5.2 and 5.3.

We obtained

$$P^*(X) = \{x_1, x_2, x_3, x_4, x_5, x_6, x_7, x_8\}$$

and

$$P_*(X) = \{x_2, x_3, x_5, x_6, x_8\}$$

Thus, the boundary region of X is,

$$B_P(X) = P^*(X) - P_*(X) = \{x_1, x_4, x_7\}$$

Equipped with these ideas, now we can define a rough set.

5.2.5 Rough set

A set S is called rough with respect to the relation P if its P-boundary region is nonempty, that is

$$B_P(X) \neq \varphi$$

Thus, if the P-boundary region of a set is empty, then the set is crisp set else the set is rough (Pawlak, 1991). The definition of approximation is depicted in Fig. 5.1.

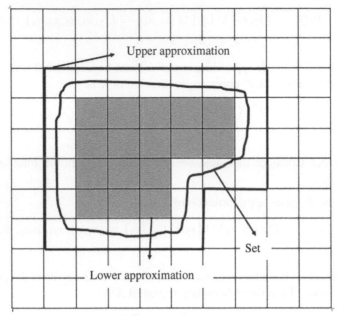

FIG. 5.1 Lower and upper approximation.

5.2.6 Fuzzy and rough sets

In the previous chapter, we consider the fuzzy set and its application along with a comparison of the crisp set. The development of the rough set from fuzzy is also interesting and necessary. The roughest can be defined with an analogous membership function, which we call the rough membership function.

5.2.7 Rough membership function

Let us consider a relation R on the set U and $R(x)$ is the equivalence class containing x. Then for a given set $X \subseteq U$, the rough membership function $\mu_X^R(x)$: $U \rightarrow [0,1]$ of X is given by

$$\mu_X^R(x) = \frac{|X \cap R(x)|}{|R(x)|}$$

where $|.|$ represents the cardinality of a set.

It is evident that $\mu_X^R(x) = 0$ if $X \cap R(x) = \emptyset$ and $\mu_X^R(x) = 1$ if $R(x) \subseteq X$. This is demonstrated in Fig. 5.2.

5.3 Reduction of attributes

As we have already mentioned, it is often necessary to maintain a concise form of the information system, and there exist data that can be removed, without altering the basic properties and more importantly the consistency of the system

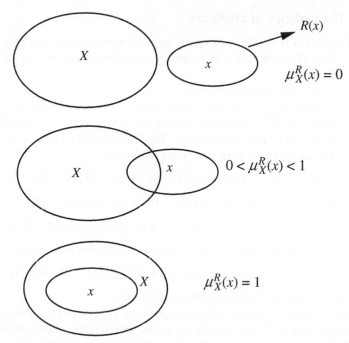

FIG. 5.2 Rough membership function of a given set X.

(Cerchiari et al., 2006). The process of reducing an information system such that the set of attributes of the reduced information system is independent and no attribute can be eliminated further without losing some information from the system, the result is known as reduct. Reducts are such minimal subsets, i.e., that do not contain any dispensable attributes. Therefore, the reduction should have the capacity to classify objects, without altering the form of representing the knowledge (Geng & Qunxiong, 2006).

5.3.1 Dispensable and indispensable feature

When a dataset is given, some features or attributes may be redundant or super-fluous and some features are essential and cannot be discarded. For a large data-set, it becomes essential to sort out the features that are superfluous to handle the dataset. A detailed discussion is given as follows.

Definition 5.7 Dispensable feature: Suppose R is an indiscernible relation with respect to the set of features S. A feature $f \in S$ is dispensable in R if

$$\text{Ind}(S) = \text{Ind}(S - \{f\})$$

5.3.2 Dependency of attributes

A set of attributes D is said to be totally dependent on another set of attributes C if all values of the attributes in D are uniquely determined by the values of attributes in C.

If this is the case, we write $C \Rightarrow D$.

To accomplish the proper understanding, we consider a similar example as shown in Table 5.1. Here, we consider three conditional attributes, FS, FE, and UI, and one decisional attribute, tenacity. We discretized the values of the attributes FS, FE, and UI into three categories as follows:

$$A_1 = [26.5 - 29), A_2 = [29 - 31.5), A_3 = [31.5 - 34]$$

$$B_1 = [5.3 - 5.83), B_2 = [5.83 - 6.36), B_3 = [6.36 - 6.90]$$

$$C_1 = [79.10 - 80.47), C_2 = [80.47 - 81.84), C_3 = [81.84 - 83.20]$$

The corresponding categories for decisional attributes are given by

$$D_1 = [10.76 - 12.68), D_2 = [12.68 - 14.60) \text{ and } D_3 = [14.60 - 16.52]$$

Suppose we get the following observation:

In this example, we observe that we cannot say $\{FS, FE, UI\} \Rightarrow T$ because of the sample x_2 and x_5. For the sample x_2 and x_5, although the values of the attributes FS, FE, and UI are the same the value of the decision attribute is not identical and thus these two data points can be considered as inconsistent. Hence, it can be inferred that the datasets represented in Table 5.2 are inconsistent.

Definition 5.8 The ratio of the number of consistent rules to the total number of rules corresponding to the set of condition attributes C and decision attributes D is called the consistency factor and is denoted by $\gamma(C, D)$.

TABLE 5.2 Dataset.

Samples	FS	FE	UI	Tenacity (T)
x_1	A_3	B_2	C_2	D_1
x_2	A_2	B_3	C_2	D_1
x_3	A_2	B_2	C_1	D_2
x_4	A_3	B_2	C_3	D_3
x_5	A_2	B_3	C_2	D_3
x_6	A_3	B_2	C_1	D_2
x_7	A_1	B_1	C_3	D_1
x_8	A_1	B_1	C_1	D_2

So, if $\gamma(C,D)=1$, we say that $C \Rightarrow D$. We say that D depends on C in a degree of k or $C \Rightarrow kD$ if $\gamma(C,D)=k$. If $k=1$, we say that D depends on C totally, else we say D, depends on C partially. In the earlier example, one can observe that the degree of dependency of tenacity on UI that is $\{UI\} \Rightarrow \{Tenacity\}$ is $\frac{3}{8}$ because only three objects x_3, x_6, and x_8 are consistent for framing the rule $\{UI\} \Rightarrow \{Tenacity\}$. This can also be achieved in the following way.

First, consider the equivalence classes of the decision attribute T.

$$\text{Ind}(T) = \{\{x_1, x_2, x_7\}, \{x_3, x_6, x_8\}, \{x_4, x_5\}\}$$

Now,

$$\text{Ind}(C = UI) = \{\{x_3, x_6, x_8\}, \{x_1, x_2, x_5\}, \{x_4, x_7\}\}$$

We need to identify those equivalence classes of UI that are totally contained in some equivalence class of D. In the earlier case we can observe that only the equivalence class $\{x_3, x_6, x_8\}$ is totally contained in the equivalence class $\{x_3, x_6, x_8\}$ of T. The cardinality of this set is 3 and hence the degree of dependency of T on UI is $\frac{3}{8}$. Similarly,

$$\text{Ind}(FE, UI) = \{\{x_1\}, \{x_2, x_5\}, \{x_3, x_6\}, \{x_4\}, \{x_7\}, \{x_8\}\}$$

Now one can observe that among the six equivalence classes in $\text{Ind}(FE, UI)$, the equivalence classes $\{x_1\}$, $\{x_3, x_6\}$, $\{x_4\}$, $\{x_7\}$, $\{x_8\}$ are totally contained in some equivalence classes of $\text{Ind}(T)$. Now the cardinality of $\{x_1\} \cup \{x_3, x_6\} \cup \{x_4\} \cup \{x_7\} \cup \{x_8\}$ is 6. So, the degree of dependency of the attribute T on (FE, UI) is $\frac{6}{8} = \frac{3}{4}$

In many cases, a data table contains one or more attributes that are not essential or superfluous. While dealing with big data it is natural that we try to remove some data or attribute from a data table without altering the basic structure of the dataset. This enables us to frame a concise decision rule. This can be achieved in the following way:

Definition 5.9 Suppose B is a subset of the set of all attributes. An attribute a in B is called dispensable in B if the following relation holds.

$$\text{Ind}(B) = \text{Ind}(B - \{a\}).$$

Otherwise, the attribute is indispensable in B.

Definition 5.10 A set of attributes B is called independent if all its attributes are indispensable.

In a large information system, we find out the dispensable or superfluous attributes and discard them. The remaining comparatively smaller system surrogates the original system. This smaller system is known as a reduct of the original system.

Definition 5.11 Let B be a set of attributes. A set B' is called a reduct of B, if B' is independent and $\text{Ind}(B) = \text{Ind}(B')$.

There exist some attributes that cannot be discarded or are indispensable. The collection of all such attributes must exist in every reduct. This collection is the core of the given set B of attributes.

Definition 5.12 Suppose B is a set of attributes. The core of B is the intersection of all reducts of B that is

$\text{Core}(B) = \bigcap \text{red}(B)$
where $\text{red}(B)$ is the reduct of B.

For better understanding, we consider the following example.

Example 5.5 We consider Table 5.3 for information that is obtained from Table 5.2 with a small modification.

In this example, we can see that

$$\text{Ind}(\text{FS}, \text{FE}, \text{UI}, \text{T}) = \{\{x_1\}, \{x_2, x_5\}, \{x_3, x_6\}, \{x_4\}\}$$

and

$$\text{Ind}(\text{FE}, \text{UI}, \text{T}) = \{\{x_1\}, \{x_2, x_5\}, \{x_3, x_6\}, \{x_4\}\}$$

This shows that, $\text{Ind}(\text{FS}, \text{FE}, \text{UI}, \text{T}) = \text{Ind}(\text{FE}, \text{UI}, \text{T})$ and hence FS is dispensable in $(\text{FS}, \text{FE}, \text{UI}, \text{T})$.

Now we consider a subset $B = \{\text{FE}, \text{UI}, \text{T}\}$ of the set of attributes $\{\text{FS}, \text{FE}, \text{UI}, \text{T}\}$.

We observe that $\text{Ind}(\text{UI}, \text{T}) = \{\{x_1, x_2, x_5\}, \{x_3, x_6\}, \{x_4\}\}$. Thus,

$$\text{Ind}(\text{FE}, \text{UI}, \text{T}) \neq \text{Ind}(\text{UI}, \text{T})$$

which implies that FE is indispensable in B.

TABLE 5.3 Dataset.

Samples	FS	FE	UI	Tenacity (T)
x_1	A_3	B_1	C_2	D_1
x_2	A_2	B_3	C_2	D_1
x_3	A_2	B_2	C_1	D_2
x_4	A_1	B_2	C_3	D_3
x_5	A_2	B_3	C_2	D_1
x_6	A_2	B_2	C_1	D_2

Now we will elaborate how to find reducts of the set B.
We have already noted that FE is indispensable in B. Now,

$$\text{Ind(FE, T)} = \{\{x_1\}, \{x_2, x_5\}, \{x_3, x_6\}, \{x_4\}\} = \text{Ind(FE, UI, T)}$$

This shows that UI is dispensable in B. Similarly, one can verify that

$$\text{Ind(FE, UI)} = \{\{x_1\}, \{x_2, x_5\}, \{x_3, x_6\}, \{x_4\}\}$$

So, T is also dispensable. Thus, to obtain a reduct of B, we must include the attribute FE in every reduct of B. In this case, we get two reducts (FE, T) and (FE, UI). Note that $\{\text{FE, T}\} \cap \{\text{FE, UI}\} = \text{FE}$. This shows that for this dataset the core of B is the attribute FE.

5.4 Application of rough set on rule generation for yarn strength prediction

In this study, the cotton dataset is used to demonstrate rule generation. The summary of the cotton dataset is presented in Table 5.4 consisting of six condition attributes and a decision attribute. A Rough Set Toolkit for Analysis of Data (ROSETTA) is used for finding decision rules.

The cotton fiber data and corresponding yarns made from the fibers in the rotor spinning system are collected from the industry. Table 5.4 shows the 108 lots of fiber and yarn datasets used in this study. The input data incorporates three essential attributes such as tensile properties, length, and mass. The FS and FE are considered tensile properties expressed in cN/tex and percentage, respectively. The UHML and length UI as length properties. The UHML is the average length of the longer group of fiber in inches and UI is defined as the mean fiber length to the UHML expressed in percentage. However, the fineness property represents the linear mass density of the fiber and yarn as FF and YF, respectively. The unit of FF is µg/inch whereas YF is calculated in the cotton count system (Ne). In this study, yarn tenacity is considered as the output parameter that is the principal component of yarn quality. The yarn tenacity is measured by the ratio of force (cN) required to break the yarn and its linear mass density (tex). The "tex" value is calculated as the mass of yarn in grams per 1 km length. The High Volume Instrument was used to measure the fiber properties such as FS, FE, UHML, UI, and FF. The FF was measured using the principle of airflow through a fire plug. The FS and FE were measured using the constant rate of extension (CRE) principle. The UHML and UI were measured using the digital fibrograph method. The yarn tenacity was measured by the universal tensile tester. The lea weight of individual yarn was measured with the help of a simple weighing balance to find out YF value. 85% of the dataset is used for the formulation of decision rules and the remaining 15% dataset is used for validation of the decision rules envisaged by the model.

TABLE 5.4 Condition attributes and decision attributes.

Cotton no.	FS	FE	UHML	UI	FF	YF	Tenacity
1	27.50	6.90	1.01	81.70	4.50	10.00	14.48
2	29.00	6.70	1.05	81.90	4.20	30.10	12.24
3	30.60	6.60	1.07	83.10	4.70	21.90	13.87
4	26.50	5.80	1.09	81.50	3.80	21.70	13.02
5	30.30	6.70	1.10	83.20	4.40	21.50	13.21
6	30.20	6.70	1.06	82.30	4.30	30.40	12.71
7	30.20	6.70	1.06	82.30	4.30	21.80	13.64
8	34.00	6.60	1.20	82.80	3.80	22.10	14.58
9	30.30	6.70	1.10	83.20	4.40	30.20	12.70
10	29.10	6.10	1.05	81.70	4.00	21.80	13.04
11	29.50	6.40	1.02	81.90	4.80	10.10	15.02
12	29.00	6.00	1.04	81.40	5.00	30.20	12.40
13	28.70	5.70	1.10	79.20	3.70	21.70	12.61
14	27.40	6.20	0.96	79.90	4.60	10.10	14.35
15	26.80	5.30	1.00	80.80	4.90	22.20	12.29
16	28.70	6.50	1.09	81.00	4.40	29.80	12.33
17	27.50	6.90	1.01	81.70	4.50	30.20	12.04
18	29.10	6.10	1.05	81.70	4.00	10.00	14.46
19	28.40	6.30	1.02	80.30	4.30	10.10	14.24
20	29.00	6.70	1.05	81.90	4.20	10.00	14.67
21	28.30	6.50	0.97	81.50	3.80	10.10	15.01
22	31.70	6.30	1.03	80.60	3.70	30.40	12.87
23	28.80	5.50	1.05	82.60	4.10	10.00	16.52
24	34.00	6.60	1.20	82.80	3.80	30.40	13.74
25	27.20	5.50	0.98	79.10	4.60	29.80	10.76
26	29.30	6.00	1.03	81.20	4.40	10.10	14.83
27	27.20	5.50	0.98	79.10	4.60	10.10	12.97
28	28.80	5.50	1.05	82.60	4.10	29.60	13.43
29	27.40	6.20	0.96	79.90	4.60	22.10	13.01
30	28.40	6.30	1.02	80.30	4.30	29.70	12.10
31	27.70	5.50	1.05	81.50	4.70	10.20	14.19
32	30.30	6.70	1.10	83.20	4.40	10.10	15.03

TABLE 5.4 Condition attributes and decision attributes—cont'd

Cotton no.	FS	FE	UHML	UI	FF	YF	Tenacity
33	30.80	6.40	1.01	81.70	3.70	29.90	13.29
34	28.90	6.00	1.07	81.10	4.60	22.40	12.88
35	28.10	6.30	1.01	80.70	3.80	29.60	12.27
36	27.50	6.90	1.01	81.70	4.50	22.10	12.79
37	28.60	5.70	1.04	82.40	4.20	21.80	13.48
38	31.70	6.30	1.03	80.60	3.70	22.60	14.05
39	29.10	6.90	1.05	83.20	4.60	10.10	14.60
40	26.70	6.90	1.04	82.60	4.80	30.00	12.51
41	28.70	6.70	1.05	81.00	3.90	10.00	14.52
42	29.20	5.30	0.98	80.00	4.50	21.80	13.19
43	28.70	5.70	1.10	79.20	3.70	30.30	11.29
44	30.60	6.60	1.07	83.10	4.70	29.90	12.89
45	28.30	6.50	0.97	81.50	3.80	30.50	12.59
46	34.00	6.60	1.20	82.80	3.80	10.00	15.82
47	29.50	6.40	1.02	81.90	4.80	30.60	12.52
48	29.30	6.00	1.03	81.20	4.40	30.10	12.76
49	28.50	6.60	1.15	80.20	3.50	29.90	11.96
50	28.30	6.50	0.97	81.50	3.80	21.70	13.86
51	29.10	6.10	1.05	81.70	4.00	30.00	12.44
52	28.10	6.70	1.03	81.70	4.50	10.20	14.25
53	28.90	6.00	1.07	81.10	4.60	30.50	12.01
54	28.70	6.20	1.02	80.70	3.80	10.10	14.92
55	29.00	6.00	1.04	81.40	5.00	22.00	13.22
56	30.80	6.40	1.01	81.70	3.70	10.00	15.58
57	30.20	6.70	1.06	82.30	4.30	10.20	15.13
58	29.50	6.40	1.02	81.90	4.80	22.00	13.56
59	29.10	6.90	1.05	83.20	4.60	30.20	12.45
60	28.60	5.70	1.04	82.40	4.20	30.00	12.11
61	28.40	6.30	1.02	80.30	4.30	21.90	12.69
62	29.30	6.00	1.03	81.20	4.40	22.00	13.75
63	29.00	6.60	1.06	80.70	3.10	9.90	15.16

Continued

TABLE 5.4 Condition attributes and decision attributes—cont'd

Cotton no.	FS	FE	UHML	UI	FF	YF	Tenacity
64	30.30	6.10	1.10	80.60	4.60	10.00	15.15
65	27.70	5.50	1.05	81.50	4.70	30.40	12.17
66	29.20	5.30	0.98	80.00	4.50	29.80	12.72
67	26.50	5.80	1.09	81.50	3.80	10.00	14.38
68	27.50	6.30	1.07	82.80	4.50	21.80	13.27
69	27.70	5.50	1.05	81.50	4.70	21.70	13.03
70	30.30	6.10	1.10	80.60	4.60	22.00	13.83
71	28.10	6.30	1.01	80.70	3.80	21.70	13.22
72	28.10	6.70	1.03	81.70	4.50	22.30	12.65
73	29.00	6.70	1.05	81.90	4.20	21.50	13.48
74	29.00	6.60	1.06	80.70	3.10	21.80	13.44
75	29.20	5.30	0.98	80.00	4.50	9.90	15.02
76	30.30	6.10	1.10	80.60	4.60	30.60	12.83
77	28.80	5.50	1.05	82.60	4.10	21.90	14.57
78	26.70	6.90	1.04	82.60	4.80	22.00	12.92
79	29.00	6.60	1.06	80.70	3.10	29.80	13.21
80	26.70	6.90	1.04	82.60	4.80	10.00	14.10
81	27.50	6.30	1.07	82.80	4.50	29.60	12.68
82	28.70	6.70	1.05	81.00	3.90	21.90	13.22
83	28.70	6.20	1.02	80.70	3.80	22.20	13.62
84	30.80	6.40	1.13	82.60	4.30	21.90	13.64
85	28.90	6.00	1.07	81.10	4.60	10.00	14.51
86	28.70	6.70	1.05	81.00	3.90	30.00	12.13
87	28.70	5.70	1.10	79.20	3.70	10.00	13.58
88	27.20	5.50	0.98	79.10	4.60	22.20	11.62
89	28.10	6.30	1.01	80.70	3.80	9.90	14.01
90	26.80	5.30	1.00	80.80	4.90	30.10	11.50
91	29.10	6.90	1.05	83.20	4.60	22.20	13.16
92	27.40	6.20	0.96	79.90	4.60	30.20	11.91
93	26.50	5.80	1.09	81.50	3.80	29.70	12.14
94	28.10	6.70	1.03	81.70	4.50	30.30	11.99

TABLE 5.4 Condition attributes and decision attributes—cont'd

Cotton no.	FS	FE	UHML	UI	FF	YF	Tenacity
95	26.80	5.30	1.00	80.80	4.90	10.10	13.60
96	30.80	6.40	1.13	82.60	4.30	10.00	14.76
97	30.80	6.40	1.01	81.70	3.70	21.90	13.88
98	31.70	6.30	1.03	80.60	3.70	10.00	14.83
99	27.50	6.30	1.07	82.80	4.50	10.10	13.70
100	28.70	6.20	1.02	80.70	3.80	29.90	12.79
101	30.60	6.60	1.07	83.10	4.70	10.10	15.19
102	28.50	6.60	1.15	80.20	3.50	21.80	12.93
103	28.70	6.50	1.09	81.00	4.40	10.20	14.62
104	29.00	6.00	1.04	81.40	5.00	10.10	14.99
105	30.80	6.40	1.13	82.60	4.30	30.00	12.96
106	28.60	5.70	1.04	82.40	4.20	10.00	15.04
107	28.70	6.50	1.09	81.00	4.40	21.60	13.44
108	28.50	6.60	1.15	80.20	3.50	10.10	14.04

5.4.1 Discretization of the dataset

The equi-frequency method is employed to discretize all 108 datasets. The number of discrete grades is set to three. The principle of discretization (Liu and Yu, 2008) is that the dimension of the attribute is small as far as possible after discretization and the lost information of attribute value is few as far as possible. For discretization of the attributes "step" of the attributes is calculated as follows:

$$Step = \frac{(maximum - minimum)}{3} \quad (5.1)$$

The ranges of each attribute were determined based on its position in either [minimum and (minimum+step)), or [(minimum+step) and (minimum +2 × step)), or [(minimum+2 × step) and maximum]. When the discretized values include the boundary values the sign of boundary brackets is used as "[ ]." But the sign of boundary brackets "[ )" indicates that the range includes the lower boundary value and the upper boundary is not included. The difference between the brackets "[ ]" and "[ )" is illustrated with the following example.

In Table 5.4, the maximum and minimum value of FF is 5.0 µg/in. and 3.1 µg/in., respectively. The step for FF is worked out to be 0.633 as per Eq. (5.1). The FF of dataset 1 is 4.5 µg/inch, which is between [(minimum +2 × step) and maximum]. Table 5.2 illustrates the discretized FF of dataset 1 as [4.37–5.0]. The brackets "[ ]" include both the boundaries, i.e., 4.37 and 5.0. The FF of dataset 2 is 4.2, which is between [(minimum + step) and (minimum + 2 × step)), and discretized to [3.73–4.37). Here, The brackets "[ )" of discretized values include the lower boundary, i.e., 3.73 but do not include the upper boundary, i.e., 4.37. Similarly, the steps of FS, FE, UHML, UI, YF, and yarn tenacity are calculated as 2.5, 0.53, 0.08, 1.36, 6.9, and 1.92, respectively. The respective steps are used to discretize the dataset as tabulated in Table 5.5.

5.4.2 Reduct operation on the cotton dataset

The most important operation of rough set theory for an optimum rule generation problem is reduct operation. The redundant attributes are eliminated to reduce the dimension of the dataset as per the attribute reduction principles. Johnson Algorithm (Johnson, 1974) is a famous approach to calculating reducts and extracting decision rules from a decision table. It always selects the most frequent attribute in the decision-making function or a row of decision-making matrices and it continues until the reducts are obtained. This algorithm considers the attribute that most often appears as the most significant one. Even though this is not true in all cases, an optimal solution is usually found. The rough set theory obtained 24 reducts data from 108 datasets along with the length of the attributes are shown in Table 5.6. The length of the reducts indicates the number of essential attributes necessary for the generation of a rule.

5.4.3 Rule generation for prediction of yarn tenacity

85% of the whole dataset, i.e., 92 datasets were used to generate decision rules for yarn tenacity prediction. The remaining 15%, i.e., 16 cotton datasets are used for testing the decision rules generated by the rough set model. From 92 datasets, only 24 combinations of attributes are approximated by the rough set model as a reduct set, which is capable of generating decision rules without losing information about the original dataset. Table 5.6 shows the reduct set obtained by rough set theory. Only 45 decision rules are generated from the reduct set as shown in Table 5.7.

ROSETTA software is used to execute indiscernibility, upper approximation, lower approximation, discretization, and reduct operation effectively for rule generation problems. Table 5.6 represents the 24 combinations of condition attributes indispensable to preserving the features of the original dataset. It is observed from Table 5.6 that the length of the reducts generally varies from 2 to 5. It can be inferred that all six attributes together are redundant to generate decision rules. In comparison with fuzzy set theory, rough set theory deals with

TABLE 5.5 Discretized dataset.

Cotton no.	FS	FE	UHML	UI	FF	YF	Tenacity
1	[26.5–29)	[6.36–6.9]	[0.96–1.04)	[80.47–81.84)	[4.37–5.00]	[9.90–16.80)	[12.68–14.60]
2	[29–31.5)	[6.36–6.9]	[1.04–1.12)	[81.84–83.20]	[3.73–4.37)	[23.70–30.60]	[10.76–12.68]
3	[29–31.5)	[6.36–6.9]	[1.04–1.12)	[81.84–83.20]	[4.37–5.00]	[16.80–23.70)	[12.68–14.60]
4	[26.5–29)	[5.3–5.83)	[1.04–1.12)	[80.47–81.84)	[3.73–4.37)	[16.80–23.70)	[12.68–14.60]
5	[29–31.5)	[6.36–6.9]	[1.04–1.12)	[81.84–83.20]	[4.37–5.00]	[16.80–23.70)	[12.68–14.60]
6	[29–31.5)	[6.36–6.9]	[1.04–1.12)	[81.84–83.20]	[3.73–4.37)	[23.70–30.60]	[12.68–14.60]
7	[29–31.5)	[6.36–6.9]	[1.04–1.12)	[81.84–83.20]	[3.73–4.37)	[16.80–23.70)	[12.68–14.60]
8	[31.5–34)	[6.36–6.9]	[1.12–1.2]	[81.84–83.20]	[3.73–4.37)	[16.80–23.70)	[12.68–14.60]
9	[29–31.5)	[6.36–6.9]	[1.04–1.12)	[81.84–83.20]	[4.37–5.00]	[23.70–30.60]	[12.68–14.60]
10	[29–31.5)	[5.83–6.36)	[1.04–1.12)	[80.47–81.84)	[3.73–4.37)	[16.80–23.70)	[12.68–14.60]
11	[29–31.5)	[6.36–6.9]	[0.96–1.04)	[81.84–83.20]	[4.37–5.00]	[9.90–16.80)	[14.6–16.52]
12	[29–31.5)	[5.83–6.36)	[1.04–1.12)	[80.47–81.84)	[4.37–5.00]	[23.70–30.60]	[10.76–12.68]
13	[26.5–29)	[5.3–5.83)	[1.04–1.12)	[79.10–80.47)	[3.1–3.73)	[16.80–23.70)	[10.76–12.68]
14	[26.5–29)	[5.83–6.36)	[0.96–1.04)	[79.10–80.47)	[4.37–5.00]	[9.90–16.80)	[12.68–14.60]
15	[26.5–29)	[5.3–5.83)	[0.96–1.04)	[80.47–81.84)	[4.37–5.00]	[16.80–23.70)	[10.76–12.68]
16	[26.5–29)	[6.36–6.9]	[1.04–1.12)	[80.47–81.84)	[4.37–5.00]	[23.70–30.60]	[10.76–12.68]
17	[26.5–29)	[6.36–6.9]	[0.96–1.04)	[80.47–81.84)	[4.37–5.00]	[23.70–30.60]	[10.76–12.68]

Continued

TABLE 5.5 Discretized dataset—cont'd

Cotton no.	FS	FE	UHML	UI	FF	YF	Tenacity
18	[29–31.5)	[5.83–6.36)	[1.04–1.12)	[80.47–81.84)	[3.73–4.37)	[9.90–16.80)	[12.68–14.60)
19	[26.5–29)	[5.83–6.36)	[0.96–1.04)	[79.10–80.47)	[3.73–4.37)	[9.90–16.80)	[12.68–14.60)
20	[29–31.5)	[6.36–6.9)	[1.04–1.12)	[81.84–83.20)	[3.73–4.37)	[9.90–16.80)	[14.6–16.52]
21	[26.5–29)	[6.36–6.9)	[0.96–1.04)	[80.47–81.84)	[3.73–4.37)	[9.90–16.80)	[14.6–16.52]
22	[31.5–34]	[5.83–6.36)	[0.96–1.04)	[80.47–81.84)	[3.1–3.73)	[23.70–30.60)	[12.68–14.60)
23	[26.5–29)	[5.3–5.83)	[1.04–1.12)	[81.84–83.20)	[3.73–4.37)	[9.90–16.80)	[14.6–16.52]
24	[31.5–34]	[6.36–6.9)	[1.12–1.2]	[81.84–83.20)	[3.73–4.37)	[23.70–30.60)	[12.68–14.60)
25	[26.5–29)	[5.3–5.83)	[0.96–1.04)	[79.10–80.47)	[4.37–5.00)	[23.70–30.60)	[10.76–12.68)
26	[29–31.5)	[5.83–6.36)	[0.96–1.04)	[80.47–81.84)	[4.37–5.00)	[9.90–16.80)	[14.6–16.52]
27	[26.5–29)	[5.3–5.83)	[0.96–1.04)	[79.10–80.47)	[4.37–5.00)	[9.90–16.80)	[12.68–14.60)
28	[26.5–29)	[5.3–5.83)	[1.04–1.12)	[81.84–83.20)	[3.73–4.37)	[23.70–30.60)	[12.68–14.60)
29	[26.5–29)	[5.83–6.36)	[0.96–1.04)	[79.10–80.47)	[4.37–5.00)	[16.80–23.70)	[12.68–14.60)
30	[26.5–29)	[5.83–6.36)	[0.96–1.04)	[79.10–80.47)	[3.73–4.37)	[23.70–30.60)	[10.76–12.68)
31	[26.5–29)	[5.3–5.83)	[1.04–1.12)	[80.47–81.84)	[4.37–5.00)	[9.90–16.80)	[12.68–14.60)
32	[29–31.5)	[6.36–6.9]	[1.04–1.12)	[81.84–83.20)	[4.37–5.00)	[9.90–16.80)	[14.6–16.52]
33	[29–31.5)	[6.36–6.9]	[0.96–1.04)	[80.47–81.84)	[3.1–3.73)	[23.70–30.60)	[12.68–14.60)
34	[26.5–29)	[5.83–6.36)	[1.04–1.12)	[80.47–81.84)	[4.37–5.00)	[16.80–23.70)	[12.68–14.60)

35	[26.5–29]	[5.83–6.36)	[0.96–1.04]	[80.47–81.84]	[3.73–4.37)	[23.70–30.60]	[10.76–12.68]
36	[26.5–29]	[6.36–6.9)	[0.96–1.04]	[80.47–81.84]	[4.37–5.00]	[16.80–23.70)	[12.68–14.60)
37	[26.5–29]	[5.3–5.83)	[1.04–1.12)	[81.84–83.20)	[3.73–4.37)	[16.80–23.70)	[12.68–14.60)
38	[31.5–34]	[5.83–6.36)	[0.96–1.04]	[80.47–81.84]	[3.1–3.73)	[16.80–23.70)	[12.68–14.60)
39	[29–31.5)	[6.36–6.9)	[1.04–1.12)	[81.84–83.20]	[4.37–5.00]	[9.90–16.80)	[14.6–16.52]
40	[26.5–29]	[6.36–6.9)	[1.04–1.12)	[81.84–83.20]	[4.37–5.00]	[23.70–30.60]	[10.76–12.68)
41	[26.5–29]	[6.36–6.9)	[1.04–1.12)	[80.47–81.84]	[3.73–4.37)	[9.90–16.80)	[12.68–14.60)
42	[29–31.5)	[5.3–5.83)	[0.96–1.04]	[79.10–80.47)	[4.37–5.00]	[16.80–23.70)	[12.68–14.60)
43	[26.5–29]	[5.3–5.83)	[1.04–1.12)	[79.10–80.47)	[3.1–3.73)	[23.70–30.60]	[10.76–12.68)
44	[29–31.5)	[6.36–6.9)	[1.04–1.12)	[81.84–83.20]	[4.37–5.00]	[23.70–30.60]	[12.68–14.60)
45	[26.5–29]	[6.36–6.9)	[0.96–1.04]	[80.47–81.84]	[3.73–4.37)	[23.70–30.60]	[10.76–12.68)
46	[31.5–34]	[6.36–6.9)	[1.12–1.2]	[81.84–83.20]	[3.73–4.37)	[9.90–16.80)	[14.6–16.52]
47	[29–31.5)	[6.36–6.9)	[0.96–1.04]	[81.84–83.20]	[4.37–5.00]	[23.70–30.60]	[10.76–12.68)
48	[29–31.5)	[5.83–6.36)	[0.96–1.04]	[80.47–81.84]	[4.37–5.00]	[23.70–30.60]	[12.68–14.60)
49	[26.5–29]	[6.36–6.9)	[1.12–1.2]	[79.10–80.47)	[3.1–3.73)	[23.70–30.60]	[10.76–12.68)
50	[26.5–29]	[6.36–6.9)	[0.96–1.04]	[80.47–81.84]	[3.73–4.37)	[16.80–23.70)	[12.68–14.60)
51	[29–31.5)	[5.83–6.36)	[1.04–1.12)	[80.47–81.84]	[3.73–4.37)	[23.70–30.60]	[10.76–12.68)
52	[26.5–29]	[6.36–6.9)	[0.96–1.04]	[80.47–81.84]	[4.37–5.00]	[9.90–16.80)	[12.68–14.60)
53	[26.5–29]	[5.83–6.36)	[1.04–1.12)	[80.47–81.84]	[4.37–5.00]	[23.70–30.60]	[10.76–12.68)

Continued

TABLE 5.5 Discretized dataset—cont'd

Cotton no.	FS	FE	UHML	UI	FF	YF	Tenacity
54	[26.5–29)	[5.83–6.36)	[0.96–1.04)	[80.47–81.84)	[3.73–4.37)	[9.90–16.80)	[14.6–16.52]
55	[29–31.5)	[5.83–6.36)	[1.04–1.12)	[80.47–81.84)	[4.37–5.00]	[16.80–23.70)	[12.68–14.60)
56	[29–31.5)	[6.36–6.9)	[0.96–1.04)	[80.47–81.84)	[3.1–3.73)	[9.90–16.80)	[14.6–16.52]
57	[29–31.5)	[6.36–6.9)	[1.04–1.12)	[81.84–83.20]	[3.73–4.37)	[9.90–16.80)	[14.6–16.52]
58	[29–31.5)	[6.36–6.9)	[0.96–1.04)	[81.84–83.20]	[4.37–5.00]	[16.80–23.70)	[12.68–14.60)
59	[29–31.5)	[6.36–6.9)	[1.04–1.12)	[81.84–83.20]	[4.37–5.00]	[23.70–30.60]	[10.76–12.68)
60	[26.5–29)	[5.3–5.83)	[1.04–1.12)	[81.84–83.20]	[3.73–4.37)	[23.70–30.60]	[10.76–12.68)
61	[26.5–29)	[5.83–6.36)	[0.96–1.04)	[79.10–80.47)	[3.73–4.37)	[16.80–23.70)	[12.68–14.60)
62	[29–31.5)	[5.83–6.36)	[0.96–1.04)	[80.47–81.84)	[4.37–5.00]	[16.80–23.70)	[12.68–14.60)
63	[29–31.5)	[6.36–6.9)	[1.04–1.12)	[80.47–81.84)	[3.1–3.73)	[9.90–16.80)	[14.6–16.52]
64	[29–31.5)	[5.83–6.36)	[1.04–1.12)	[80.47–81.84)	[4.37–5.00]	[9.90–16.80)	[14.6–16.52]
65	[26.5–29)	[5.3–5.83)	[1.04–1.12)	[80.47–81.84)	[4.37–5.00]	[23.70–30.60]	[10.76–12.68)
66	[29–31.5)	[5.3–5.83)	[0.96–1.04)	[79.10–80.47)	[4.37–5.00]	[23.70–30.60]	[12.68–14.60)
67	[26.5–29)	[5.3–5.83)	[1.04–1.12)	[80.47–81.84)	[3.73–4.37)	[9.90–16.80)	[12.68–14.60)
68	[26.5–29)	[5.83–6.36)	[1.04–1.12)	[81.84–83.20]	[4.37–5.00]	[16.80–23.70)	[12.68–14.60)
69	[26.5–29)	[5.3–5.83)	[1.04–1.12)	[80.47–81.84)	[4.37–5.00]	[16.80–23.70)	[12.68–14.60)
70	[29–31.5)	[5.83–6.36)	[1.04–1.12)	[80.47–81.84)	[4.37–5.00]	[16.80–23.70)	[12.68–14.60)

71	[26.5–29)	[5.83–6.36)	[0.96–1.04)	[80.47–81.84)	[3.73–4.37]	[16.80–23.70)	[12.68–14.60)
72	[26.5–29)	[6.36–6.9]	[0.96–1.04)	[80.47–81.84)	[4.37–5.00]	[16.80–23.70)	[10.76–12.68)
73	[29–31.5)	[6.36–6.9]	[1.04–1.12)	[81.84–83.20]	[3.73–4.37]	[16.80–23.70)	[12.68–14.60)
74	[29–31.5)	[6.36–6.9]	[1.04–1.12)	[80.47–81.84)	[3.1–3.73)	[16.80–23.70)	[12.68–14.60)
75	[29–31.5)	[5.3–5.83)	[0.96–1.04)	[79.10–80.47]	[4.37–5.00]	[9.90–16.80)	[14.6–16.52]
76	[29–31.5)	[5.83–6.36)	[1.04–1.12)	[80.47–81.84)	[4.37–5.00]	[23.70–30.60]	[12.68–14.60)
77	[5.3–5.83)		[1.04–1.12)	[81.84–83.20]	[3.73–4.37]	[16.80–23.70)	[12.68–14.60)
78	[26.5–29)	[6.36–6.9]	[1.04–1.12)	[81.84–83.20]	[4.37–5.00]	[16.80–23.70)	[12.68–14.60)
79	[29–31.5)	[6.36–6.9]	[1.04–1.12)	[80.47–81.84)	[3.1–3.73)	[23.70–30.60]	[12.68–14.60)
80	[26.5–29)	[6.36–6.9]	[1.04–1.12)	[81.84–83.20]	[4.37–5.00]	[9.90–16.80)	[12.68–14.60)
81	[26.5–29)	[5.83–6.36)	[1.04–1.12)	[81.84–83.20]	[4.37–5.00]	[23.70–30.60]	[12.68–14.60)
82	[26.5–29)	[6.36–6.9]	[1.04–1.12)	[80.47–81.84)	[3.73–4.37]	[16.80–23.70)	[12.68–14.60)
83	[26.5–29)	[5.83–6.36]	[0.96–1.04)	[80.47–81.84)	[3.73–4.37]	[16.80–23.70)	[12.68–14.60)
84	[29–31.5)	[6.36–6.9]	[1.12–1.2]	[81.84–83.20]	[3.73–4.37]	[16.80–23.70)	[12.68–14.60)
85	[26.5–29)	[5.83–6.36)	[1.04–1.12)	[80.47–81.84)	[4.37–5.00]	[9.90–16.80)	[12.68–14.60)
86	[26.5–29)	[6.36–6.9]	[1.04–1.12)	[80.47–81.84)	[3.73–4.37]	[23.70–30.60]	[10.76–12.68)
87	[26.5–29)	[5.3–5.83)	[1.04–1.12)	[79.10–80.47]	[3.1–3.73)	[9.90–16.80)	[12.68–14.60)
88	[26.5–29)	[5.3–5.83)	[0.96–1.04)	[79.10–80.47]	[4.37–5.00]	[16.80–23.70)	[10.76–12.68)
89	[26.5–29)	[5.83–6.36)	[0.96–1.04)	[80.47–81.84)	[3.73–4.37]	[9.90–16.80)	[12.68–14.60)
90	[26.5–29)	[5.3–5.83)	[0.96–1.04)	[80.47–81.84)	[4.37–5.00]	[23.70–30.60]	[10.76–12.68)

Continued

TABLE 5.5 Discretized dataset—cont'd

Cotton no.	FS	FE	UHML	UI	FF	YF	Tenacity
91	[29–31.5)	[6.36–6.9]	[1.04–1.12)	[81.84–83.20]	[4.37–5.00]	[16.80–23.70)	[12.68–14.60)
92	[26.5–29)	[5.83–6.36)	[0.96–1.04)	[79.10–80.47)	[4.37–5.00]	[23.70–30.60]	[10.76–12.68)
93	[26.5–29)	[5.3–5.83)	[1.04–1.12)	[80.47–81.84)	[3.73–4.37)	[23.70–30.60]	[10.76–12.68)
94	[26.5–29)	[6.36–6.9]	[0.96–1.04)	[80.47–81.84)	[4.37–5.00]	[23.70–30.60]	[10.76–12.68)
95	[26.5–29)	[5.3–5.83)	[0.96–1.04)	[80.47–81.84)	[4.37–5.00]	[9.90–16.80)	[12.68–14.60)
96	[29–31.5)	[6.36–6.9]	[1.12–1.2]	[81.84–83.20]	[3.73–4.37)	[9.90–16.80)	[14.6–16.52)
97	[29–31.5)	[6.36–6.9]	[0.96–1.04)	[80.47–81.84)	[3.1–3.73)	[16.80–23.70)	[12.68–14.60)
98	[31.5–34)	[5.83–6.36)	[0.96–1.04)	[80.47–81.84)	[3.1–3.73)	[9.90–16.80)	[14.6–16.52)
99	[26.5–29)	[5.83–6.36)	[1.04–1.12)	[81.84–83.20]	[4.37–5.00]	[9.90–16.80)	[12.68–14.60)
100	[26.5–29)	[5.83–6.36)	[0.96–1.04)	[80.47–81.84)	[3.73–4.37)	[23.70–30.60]	[12.68–14.60)
101	[29–31.5)	[6.36–6.9]	[1.04–1.12)	[81.84–83.20]	[4.37–5.00]	[9.90–16.80)	[14.6–16.52)
102	[26.5–29)	[6.36–6.9]	[1.12–1.2]	[79.10–80.47)	[3.1–3.73)	[16.80–23.70)	[12.68–14.60)
103	[26.5–29)	[6.36–6.9]	[1.04–1.12)	[80.47–81.84)	[4.37–5.00]	[9.90–16.80)	[14.6–16.52)
104	[29–31.5)	[5.83–6.36)	[1.04–1.12)	[80.47–81.84)	[4.37–5.00]	[9.90–16.80)	[14.6–16.52)
105	[29–31.5)	[6.36–6.9]	[1.12–1.2]	[81.84–83.20]	[3.73–4.37)	[23.70–30.60]	[12.68–14.60)
106	[26.5–29)	[6.36–6.9]	[1.04–1.12)	[81.84–83.20]	[3.73–4.37)	[9.90–16.80)	[14.6–16.52)
107	[26.5–29)	[6.36–6.9]	[1.04–1.12)	[80.47–81.84)	[4.37–5.00]	[16.80–23.70)	[12.68–14.60)
108	[26.5–29)	[6.36–6.9]	[1.12–1.2]	[79.10–80.47)	[3.1–3.73)	[9.90–16.80)	[12.68–14.60)

TABLE 5.6 Reducts.

Sl. no.	Reduct attributes	Length
1	[FS, UHML, FF, YF]	4
2	[FS, UHML, UI, YF]	4
3	[FS, YF]	2
4	[FF, YF]	2
5	[FS, FE, YF]	3
6	[FS, FE, UI, YF]	4
7	[FS, UI, YF]	3
8	[FS, FE, UHML, YF]	4
9	[FS, FE, FF, YF]	4
10	[FE, UHML, FF, YF]	4
11	[FE, UI, YF]	3
12	[FS, UHML, YF]	3
13	[FE, YF]	2
14	[FS, UHML, FF]	3
15	[FS, FE, UHML, UI, YF]	5
16	[FS, FE, UHML, FF, YF]	5
17	[UI, YF]	2
18	[UHML, UI, FF, YF]	4
19	[UHML, UI, YF]	3
20	[FS, FF, YF]	3
21	[FE, UI]	2
22	[UHML, YF]	2
23	[FE, UHML, UI, FF, YF]	5
24	[FE, UHML, YF]	3

the indiscernibility among data, which refers to the granularity of knowledge, while fuzzy sets theory handles the ill-definition of the boundary of a class through a continuous generalization of membership functions. The rough set theory obtained only 45 rules to predict the cotton yarn tenacity in this study. But if we consider fuzzy set theory for this application, a cotton fire dataset with six attributes each having three linguistic variables produces an unmanageable number of rules, i.e., 36 equal to 729 to predict yarn tenacity. The reduct set is

TABLE 5.7 Decision rules.

Rule no.	IF	THEN
1	FS([26.5–29)) AND UHML([0.96–1.04)) AND FF([4.37–5]) AND YF([9.9–16.80))	Tenacity ([12.68–14.60))
2	FS([29–31.5)) AND UHML([1.04–1.12)) AND FF([4.37–5]) AND YF([23.70–30.6))	Tenacity ([12.68–14.60))
3	FS([29–31.5)) AND UHML([1.04–1.12)) AND UI([81.84–83.20]) AND YF([23.70–30.6])	Tenacity ([12.68–14.60))
4	FS([26.5–29)) AND UHML([1.04–1.12)) AND UI([80.47–81.84)) AND YF([23.70–30.6))	Tenacity ([10.76–12.68))
5	FS([29–31.5)) AND YF([16.80–23.70))	Tenacity ([12.68–14.60))
6	FS([31.5–34]) AND YF([16.80–23.70))	Tenacity ([12.68–14.60))
7	FS([31.5–34]) AND YF([23.70–30.6])	Tenacity ([12.68–14.60))
8	FS([31.5–34]) AND YF([9.9–16.80))	Tenacity ([14.60–16.52])
9	FF([3.73–4.37)) AND YF([16.80–23.70))	Tenacity ([12.68–14.60))
10	FS([29–31.5)) AND FE([6.36–6.9]) AND YF([9.9–16.80))	Tenacity ([14.60–16.52])
11	FS([26.5–29)) AND FE([6.36–6.9]) AND YF([23.70–30.6])	Tenacity ([10.76–12.68))
12	FS([29–31.5)) AND FE([5.3–5.83)) AND YF([23.70–30.6])	Tenacity ([12.68–14.60))
13	FS([29–31.5)) AND FE([5.3–5.83)) AND YF([9.9–16.80))	Tenacity ([14.60–16.52])
14	FS([26.5–29)) AND FE([5.3–5.83)) AND UI([79.10–80.47)) AND YF([16.80–23.70))	Tenacity ([10.76–12.68))
15	FS([26.5–29)) AND FE([6.36–6.9]) AND UI([81.84–83.20]) AND YF([9.9–16.80))	Tenacity ([12.68–14.60))
16	FS([26.5–29)) AND UI([79.10–80.47)) AND YF([9.9–16.80))	Tenacity ([12.68–14.60))
17	FS([26.5–29)) AND UI([79.10–80.47)) AND YF([23.70–30.6))	Tenacity ([10.76–12.68))
18	FS([26.5–29)) AND FE([5.3–5.83)) AND UHML([0.96–1.04)) AND YF([16.80–23.70))	Tenacity ([10.76–12.68))

TABLE 5.7 Decision rules—cont'd

Rule no.	IF	THEN
19	FS([29–31.5)) AND FE([5.83–6.36)) AND UHML([0.96–1.04)) AND YF([23.70–30.6])	Tenacity ([12.68–14.60))
20	FS([26.5–29)) AND FE([5.83–6.36)) AND UHML([1.04–1.12)) AND YF([9.9–16.80))	Tenacity ([12.68–14.60))
21	FS([29–31.5)) AND FE([5.83–6.36)) AND FF([3.73–4.37)) AND YF([9.9–16.80))	Tenacity ([12.68–14.60))
22	FS([29–31.5)) AND FE([5.83–6.36)) AND FF([3.73–4.37)) AND YF([23.70–30.6])	Tenacity ([10.76–12.68))
23	FE([6.36–6.9]) AND UHML([0.96–1.04)) AND FF([3.73–4.37)) AND YF([9.9–16.80))	Tenacity ([14.60–16.52])
24	FE([5.3–5.83)) AND UI([81.84–83.20]) AND YF([9.9–16.80))	Tenacity ([14.60–16.52])
25	FE([5.3–5.83)) AND UI([81.84–83.20]) AND YF([23.70–30.6])	Tenacity ([10.76–12.68))
26	FE([5.3–5.83)) AND UI([80.47–81.84)) AND YF([9.9–16.80))	Tenacity ([12.68–14.60))
27	FE([5.3–5.83)) AND UI([80.47–81.84)) AND YF([23.70–30.6])	Tenacity ([10.76–12.68))
28	FS([29–31.5)) AND UHML([0.96–1.04)) AND YF([9.9–16.80))	Tenacity ([14.60–16.52])
29	FS([29–31.5)) AND UHML([1.12–1.2]) AND YF([23.70–30.6])	Tenacity ([12.68–14.60))
30	FS([26.5–29)) AND UHML([1.12–1.2]) AND YF([9.9–16.80))	Tenacity ([12.68–14.60))
31	FE([5.83–6.36)) AND YF([16.80–23.70))	Tenacity ([12.68–14.60))
32	FS([29–31.5)) AND UHML([0.96–1.04)) AND FF([3.1–3.73))	Tenacity ([12.68–14.60))
33	FS([26.5–29)) AND FE([5.83–6.36)) AND UHML([0.96–1.04)) AND UI([80.47–81.84)) AND YF([23.70–30.6])	Tenacity ([12.68–14.60))
34	FS([26.5–29)) AND FE([5.83–6.36)) AND UHML([0.96–1.04)) AND UI([80.47–81.84)) AND YF([9.9–16.80))	Tenacity ([14.60–16.52])
35	FS([26.5–29)) AND FE([6.36–6.9]) AND UHML([0.96–1.04)) AND FF([4.37–5]) AND YF([16.80–23.70))	Tenacity ([10.76–12.68))

Continued

TABLE 5.7 Decision rules—cont'd

Rule no.	IF	THEN
36	UI([81.84–83.20]) AND YF([16.80–23.70))	Tenacity ([12.68–14.60))
37	UHML([1.04–1.12)) AND UI([80.47–81.84)) AND FF([3.73–4.37)) AND YF([9.9–16.80))	Tenacity ([12.68–14.60))
38	UHML([0.96–1.04)) AND UI([81.84–83.20]) AND YF([23.70–30.6))	Tenacity ([10.76–12.68))
39	UHML([1.04–1.12)) AND UI([80.47–81.84)) AND YF([16.80–23.70))	Tenacity ([12.68–14.60))
40	FS([29–31.5)) AND FF([4.37–5]) AND YF([9.9–16.80))	Tenacity ([14.60–16.52])
41	FS([29–31.5)) AND FF([3.1–3.73)) AND YF([23.70–30.6))	Tenacity ([12.68–14.60))
42	FE([5.83–6.36)) AND UI([81.84–83.20])	Tenacity ([12.68–14.60))
43	UHML([1.12–1.2]) AND YF([16.80–23.70))	Tenacity ([12.68–14.60))
44	FE([6.36–6.9]) AND UHML([1.04–1.12)) AND UI([80.47–81.84)) AND FF([4.37–5]) AND YF([9.9–16.80))	Tenacity ([14.60–16.52])
45	FE([6.36–6.9]) AND UHML([1.04–1.12)) AND YF([16.80–23.70))	Tenacity ([12.68–14.60))

used to formulate the decision rules to predict yarn tenacity as shown in Table 5.7. The first rule denotes that IF "FS is between 26.5 and 29" AND "UHML is between 0.96 and 1.04" AND "FF is between 4.37 and 5.00" AND "YF is between 9.90 and 16.80" THEN "yarn tenacity is between 12.68 and 14.60." In the 1st rule, only four attributes such as FS, UHML, FF, and YF were employed to predict the range of yarn tenacity. Thus rough set approach can extract the approximate decision rules from the dataset, which can be useful in realizing the acceptable solution of yarn tenacity. Cotton is a natural fire with huge variability; hence, it is always difficult to establish an exact mathematical model for predicting actual yarn tenacity from the raw material parameters. The rough set theory turns out to be the most suitable approach in such a state of affairs where accuracy is considered to be secondary and we are primarily interested in acceptable solutions.

Table 5.8 demonstrates the validation of obtained decision rules using the 15% cotton dataset of the 108 dataset. The attributes of 16 cotton datasets

TABLE 5.8 Validation.

Sl. no.	FS	FE	UHML	UI	FF	YF	Actual tenacity	Predicted tenacity
1	[26.5–29]	[5.30–5.83]	[1.04–1.12]	[80.47–81.36]	[3.73–4.37]	[16.80–23.70]	[12.68–14.60]	[12.68–14.60]
2	[29–31.5]	[6.36–6.90]	[1.04–1.12]	[81.36–83.20]	[3.73–4.37]	[23.70–30.6]	[12.68–14.60]	[12.68–14.60]
3	[29–31.5]	[6.36–6.90]	[1.04–1.12]	[81.36–83.20]	[4.37–5.00]	[23.70–30.60]	[12.68–14.60]	[12.68–14.60]
4	[26.5–29]	[5.30–5.83]	[1.04–1.12]	[79.10–80.47]	[3.10–3.73]	[16.80–23.70]	[10.76–12.68]	[12.68–14.60]
5	[29–31.5]	[6.36–6.90]	[1.04–1.12]	[81.36–83.20]	[3.73–4.37]	[9.90–16.80]	[14.60–16.52]	[14.60–16.52]
6	[31.5–34]	[5.83–6.36]	[0.96–1.04]	[80.47–81.36]	[3.10–3.73]	[23.70–30.60]	[12.68–14.60]	[12.68–14.60]
7	[29–31.5]	[6.36–6.90]	[0.96–1.04]	[80.47–81.36]	[3.10–3.73]	[23.70–30.60]	[12.68–14.60]	[12.68–14.60]
8	[26.5–29]	[5.83–6.36]	[1.04–1.12]	[80.47–81.36]	[4.37–5.00]	[16.80–23.70]	[12.68–14.60]	[12.68–14.60]
9	[26.5–29]	[6.36–6.90]	[1.04–1.12]	[80.47–81.36]	[3.73–4.37]	[9.90–16.80]	[12.68–14.60]	[12.68–14.60]
10	[29–31.5]	[6.36–6.90]	[0.96–1.04]	[81.36–83.20]	[4.37–5.00]	[23.70–30.60]	[10.76–12.68]	[12.68–14.60]
11	[26.5–29]	[6.36–6.90]	[0.96–1.04]	[80.47–81.36]	[4.37–5.00]	[9.90–16.80]	[12.68–14.60]	[12.68–14.60]
12	[29–31.5]	[6.36–6.90]	[0.96–1.04]	[80.47–81.36]	[3.10–3.73]	[9.90–16.80]	[14.60–16.52]	[14.60–16.52]
13	[29–31.5]	[6.36–6.90]	[0.96–1.04]	[81.36–83.20]	[4.37–5.00]	[16.80–23.70]	[12.68–14.60]	[12.68–14.60]
14	[26.5–29]	[5.83–6.36]	[0.96–1.04]	[79.10–80.47]	[3.73–4.37]	[16.80–23.70]	[12.68–14.60]	[12.68–14.60]
15	[29–31.5]	[6.36–6.90]	[1.04–1.12]	[80.47–81.36]	[3.10–3.73]	[9.90–16.80]	[14.60–16.52]	[14.60–16.52]
16	[29–31.5]	[5.30–5.83]	[0.96–1.04]	[79.10–80.47]	[4.37–5.00]	[23.70–30.60]	[12.68–14.60]	[12.68–14.60]

are used to predict yarn tenacity using decision rules generated by a rough set model. It is observed that the proposed rough set model effectively predicted the range of yarn tenacity for 14 out of 16 datasets, which are computed with 88% accuracy. Hence, the validation result shows that the proposed model has sufficient predictive power to support the textile domain.

5.5 Summary

The chapter serves as a comprehensive exploration of a mathematical framework designed to grapple with uncertainty and imprecision in data. Beginning with fundamental definitions such as lower and upper approximations, the chapter delves into the core principles of discernibility and boundary regions within decision tables. Pawlak's foundational model, featuring discernibility matrices and rough set approximations, is thoroughly elucidated, providing a solid theoretical foundation. The chapter also discusses about reduct and rule generation process for prediction. The practical relevance of rough set theory is highlighted through its applications in the textile domain. Overall, the chapter not only equips readers with a deep understanding of rough set theory but also encourages them to consider its practical implications and future research directions.

References

Cerchiari, S. C., Teurya, A., Pinto, J. P., Torres, G. L., Sauer, L., & Zorzate, E. H. (2006). Data mining in distribution consumer database using rough sets and self-organizing maps. In *Proceedings of the 2006 IEEE power systems conference and exposition*. NJ, USA: IEEE Press.

Geng, Z., & Qunxiong, Z. (2006). A new rough set-based heuristic algorithm for attribute reduct, Dalian, China. *6th World congress on intelligent control and automation* (pp. 3085–3089). IEEE.

Johnson, D. S. (1974). Approximation algorithms for combinatorial problems. *Journal of Computer and System Sciences*, *9*(3), 256–278.

Liu, G., & Yu, W. (2008). The application of rough set theory in worsted roving procedure. In K. Elleithy (Ed.), *Innovations and advanced techniques in systems, computing sciences and software engineering* (pp. 309–312). Dordrecht: Springer.

Pawlak, Z. (1982). Rough sets. *International Journal of Computer and Information Sciences*, *11*, 341–356.

Pawlak, Z. (1991). *Rough sets: Thoretical aspects of reasoning about data*. Boston, MA: Kluwer Academic.

Wu, C., Yue, Y., Li, M., & Adjei, O. (2004). The rough set theory and applications. *Engineering Computations*, *21*(5), 488–511.

Zadeh, L. A. (1965). Fuzzy sets. *Information and Control*, *8*(3), 338–353.

Chapter 6

Nature-inspired optimization algorithms

6.1 Introduction

The purpose of optimization is to obtain the best possible solution or achieve the best performance under certain constraints. A general problem of optimization involves finding the maximum or minimum of a function, known as the objective function, subject to a set of constraints that need to be satisfied. In the realm of optimization, humankind has always sought inspiration from the natural phenomena. Whether it is the coordinated behavior of bird flocks or fish schools, the strategic foraging of ants, or even the annealing process in metallurgy, nature's solutions have consistently inspired and guided human innovation. The invention of high-speed computational systems has greatly enhanced the development of nature-inspired optimization algorithms, which have opened new frontiers in artificial intelligence by challenging the conventional approaches in tackling complex real-world problems.

This chapter delves into the fascinating realm of nature-inspired optimization, a field of study that mimics nature's brilliance. The basic concepts of some popularly known nature-inspired optimization algorithms such as genetic algorithm, particle swarm optimization, and ant colony optimization are discussed along with their potential applications in textile engineering. These algorithms have established remarkable success in various domains such as engineering, finance, biology, and data science, among others.

6.2 Genetic algorithm (GA)

GA is a population-based probabilistic optimization technique. It mimics Darwin's principle of natural selection based on the "survival of the fittest." Prof. John Holland (1975) of the University of Michigan, Ann Arbor was the first to propose the concept of GA. Thereafter, a number of scientists, namely Goldberg (1989), Davis (1991), and Deb (2005), have contributed a lot in developing this subject. The main operations of GA are explained in the following section.

Artificial Intelligence in Textile Engineering. https://doi.org/10.1016/B978-0-443-15395-2.00002-8

6.2.1 Main operations of GA

6.2.1.1 Representation

GA starts with a population of randomly generated initial solutions of size N that are represented by binary strings. A binary string is analogous to a biological chromosome and each bit of the string is nothing but a gene value. The optimization problem is defined in the real world and the encoding or representation of the variables is done in the binary world as displayed in Fig. 6.1 (Ghosh et al., 2019). The operators such as crossover and mutation are applied in the binary world. By means of decoding or transverse representation, we go back to the real world where fitness evaluation and reproduction operations are performed.

6.2.1.2 Fitness evaluation

As GA is based on the concept of "survival of the fittest," therefore it can able to solve only a maximization problem. Hence, the minimization of a function $f(x)$ has to be converted into a suitable fitness function either by maximizing $-f(x)$, maximizing $\frac{1}{f(x)}$ for $f(x) \neq 0$, or maximizing $\frac{1}{1+f(x)}$ for $f(x) \neq -1$, and so on. The fitness values or the values of the fitness function are evaluated for each string of the population. To determine the fitness value of a binary string, the real values of the variables are to be estimated first. A binary-coded string can be converted into real values of variables by means of a linear mapping using the following expression:

$$x_i = x_{i(\min)} + \frac{x_{i(\max)} - x_{i(\min)}}{2^{l_i} - 1} \times D_i \tag{6.1}$$

where x_i is the real value, $x_{i(\max)}$ and $x_{i(\min)}$ are the maximum and minimum values, l_i is the string length, D_i is the decoded value corresponding to the ith variable.

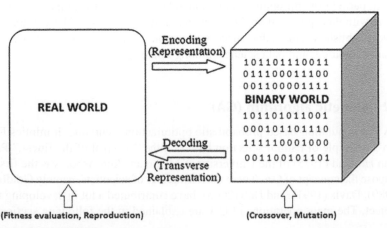

FIG. 6.1 Representation.

6.2.1.3 Reproduction

All the strings in a population may not be equally good with respect to their fitness values. The reproduction operator is utilized to choose the good strings using their fitness values. It forms a mating pool consisting of good solutions probabilistically. It thus turns out that the mating pool may contain multiple copies of a particular good solution. The size of the mating pool is kept equal to that of the initial population. The average fitness of the mating pool is expected to be higher than the average fitness of the population before reproduction. There exists a number of reproduction schemes, namely roulette-wheel selection, tournament selection, and ranking selection, which are briefly explained the following section (Ghosh et al., 2019).

Roulette wheel selection

In roulette-wheel selection, the probability of a string for being selected in the mating pool depends upon its fitness value. A string with a higher fitness value has a greater probability of being selected in the mating pool. The working principle of roulette-wheel selection is explained with the help of Fig. 6.2. Let us suppose that there are only three strings A, B, and C in the population with their fitness values 1, 2, and 3, respectively. A summation of the fitness values turns out to be 6. Hence, the selection probabilities of strings A, B, and C in the mating pool are 0.17 (1/6), 0.33 (2/6), and 0.5 (3/6), respectively, which gives the cumulative probability equals to 1. Now we design a roulette-wheel whose total area is 1 and assign the three strings to the wheel in such a way that the area shared by each string is proportional to its selection probability. Thus, it is

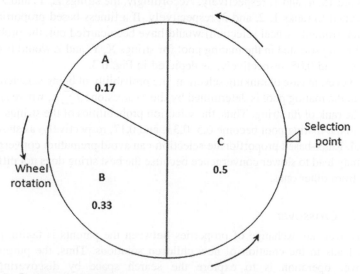

FIG. 6.2 Roulette wheel selection.

evident from Fig. 6.2 that string C shares the highest area (50%) and string A shares the lowest area (17%) in the roulette-wheel. As there are three strings, the wheel is rotated three times and each time a fixed selection point indicates the winning string. Let us assume that string C wins twice and string B wins once. Hence, string C is copied twice and string B is copied once in the mating pool, whereas string A being weaker is eliminated.

Tournament selection

In this method, a tournament is played by selecting a small number of strings of size n at random from the population of size N and the best one is determined on the basis of fitness value. The size n is kept much smaller than N. After a tournament is played, the best string is getting copied into the mating pool and then all n strings are returned to the population. Hence, only one string is copied per tournament. Subsequently, N number of tournaments are to be played to ensure the size of mating pool equals to N.

Ranking selection

In ranking selection method, firstly the strings are arranged in an ascending order of their fitness values. The strings with the lowest fitness get rank 1, the next lowest get rank 2, and so on, until the best fitness receives rank N (number of population). In the next step, a proportionate selection scheme based on the assigned rank is adopted. The working principle of ranking selection is explained with the help of the following example. Let us suppose that there are only three strings X, Y, and Z in the population whose fitness values are 15, 4, and 1, respectively. Accordingly, the strings Z, Y, and X will be assigned to ranks 1, 2, and 3, respectively. If a fitness-based proportionate selection (roulette-wheel selection) would have been carried out, the probabilities of being selected in the mating pool for strings X, Y, and Z would become 0.75, 0.2, and 0.05, respectively, as depicted in Fig. 6.3.

However, in case of ranking selection, the probability of being selected for a string in the mating pool is determined by the expression $r_i/\sum r_i$, where r_i indicates the rank of ith string. Thus, the selection probabilities of the strings X, Y, and Z in the mating pool become 0.5, 0.33, and 0.17, respectively, as shown in Fig. 6.4. Rank-based proportionate selection can avoid premature convergence but it may lead to slower convergence because the best string does not differ so much from other ones.

6.2.1.4 Crossover

In crossover, an exchange of properties between the parents is taking place, which leads to the creation of new children solutions. Thus, the purpose of crossover operation is to explore the search space by discovering the promising area.

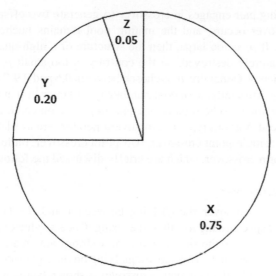

FIG. 6.3   Fitness-based proportionate selection.

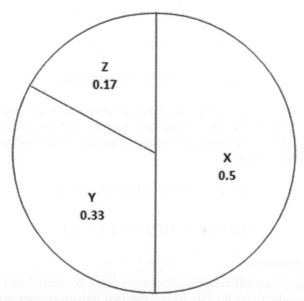

FIG. 6.4   Rank-based proportionate selection.

To begin with the crossover, firstly the mating pairs (each pair consists of two parent strings) are formed from the mating pool. If the population size is N, then $N/2$ mating pairs are formed. The decision for a mating pair to undergo crossover depends on the crossover probability (p_c). To determine crossover participation, a coin is tossed with a probability (p_c) of landing head. If head appears, a mating pair engages in crossover to generate two offspring. Otherwise, no crossover occurs, and the mating pool remains unchanged within the population. If p_c is too large, then the structure of a high-quality solution could be prematurely destroyed, on the contrary, a too small p_c reduces the searching efficiency. Generally, p_c is chosen between 0.6 and 0.9. To implement the coin tossing artificially, a random number lying between 0 and 1 is generated, and if it is found to be smaller or equal to p_c, the outcome is considered as success or head. Various types of crossover operators are available in the literature, such as single-point crossover, two-point crossover, multi-point crossover, and uniform crossover, which are briefly discussed the following section.

Single-point crossover

In this scheme, a crossover site (j) lying between 1 and ($l-1$) is randomly selected where l specifies the length of the string. Once a value of j is obtained, the strings are crossed at the site j and two new strings are created. In general, crossing is done between the two substrings lying on the right side of the crossover site. An example of single point crossover is shown below where the first parent string is represented by all ones and the second parent string is represented by all zeros.

$$1\ 1\ 1\ 1\ 1\ 1\ 1\ 1\ 1\ |1\ 1\ 1\ 1\ 1\ 1$$

$$0\ 0\ 0\ 0\ 0\ 0\ 0\ 0\ 0\ 0\ |0\ 0\ 0\ 0\ 0\ 0$$

Here the length of the string is 16. Thus, a number between 1 and 15 is created at random to choose the crossing site. Suppose that the obtained random number is 10. Hence, we cross the strings at the site 10 to produce two new strings as shown as follows:

$$1\ 1\ 1\ 1\ 1\ 1\ 1\ 1\ 1\ 1\ |0\ 0\ 0\ 0\ 0\ 0$$

$$0\ 0\ 0\ 0\ 0\ 0\ 0\ 0\ 0\ 0\ |1\ 1\ 1\ 1\ 1\ 1$$

Two-point crossover

In this scheme, two different crossover sites lying between 1 and ($l-1$) are selected at random and the bits inside the two crossover sites are swapped between the parent strings. The parent strings participating in crossover are shown below where random points 4 and 9 are chosen as two crossing sites.

$$1\ 1\ 1\ 1\ |1\ 1\ 1\ 1\ 1\ |1\ 1\ 1\ 1\ 1\ 1$$

$$0\ 0\ 0\ 0\ |0\ 0\ 0\ 0\ 0\ |0\ 0\ 0\ 0\ 0\ 0$$

The crossover is performed by swapping the bits inside 4 and 9 between the parent strings. Two children strings thus produced due to the two-point crossover are given as follows:

1 1 1 1|0 0 0 0 0|1 1 1 1 1 1

0 0 0 0| 1 1 1 1 1|0 0 0 0 0 0 0

Multi-point crossover

In this scheme, multiple numbers (more than two) of crossover sites are selected at random along the length of the string. The bits lying inside the alternate pairs of crossover sites are swapped between the parent strings. The parent strings are shown below where random points 3, 8, and 14 are chosen as three crossover sites.

1 1 1|1 1 1 1 1 |1 1 1 1 1 1|1 1

0 0 0|0 0 0 0 0 |0 0 0 0 0 0 |0 0

The bits lying inside the crossover sites 3 and 8; and the bits lying right side of crossover site 14 are swapped between the parent strings while the remaining bits are kept unaltered. The generated children strings thus produced due to the multi-point crossover are shown as follows:

1 1 1| 0 0 0 0 0|1 1 1 1 1| 0 0

0 0 0|1 1 1 1 1 |0 0 0 0 0 0|1 1

If five crossover sites are selected at random are 2, 5, 7, 11, and 13, then the parent strings will look as follows:

1 1| 1 1 1| 1 1| 1 1 1 1| 1 1| 1 1 1

0 0| 0 0 0| 0 0| 0 0 0 0| 0 0| 0 0 0

In this case the bits lying inside the crossover sites 2 and 5; 7 and 11; and the right side of 13 are swapped between the parent strings by keeping the other bits unaltered. Thus, the children strings produced after crossover will look as follows:

1 1| 0 0 0| 1 1| 0 0 0 0| 1 1| 0 0 0

0 0| 1 1 1| 0 0| 1 1 1 1| 0 0| 1 1 1

Uniform crossover

In this scheme, at each bit position of the parent strings, a coin is tossed with a probability of appearing head $= 0.5$ to determine whether there will be swapping of the bits. If head (H) appears at a bit position, there will be swapping of the bits between the parent strings. On the other hand, if the outcome is tail (T), they will remain unaltered. To explain the principle of uniform crossover, let us take the same parent strings like previous cases as follows:

$$1\ 1\ 1\ 1\ 1\ 1\ 1\ 1\ 1\ 1\ 1\ 1\ 1\ 1\ 1\ 1$$

$$0\ 0\ 0\ 0\ 0\ 0\ 0\ 0\ 0\ 0\ 0\ 0\ 0\ 0\ 0\ 0$$

Suppose that heads appear at 3rd, 6th, 7th, 10th, 11th, and 16th bit positions. After swapping of the bits between parent strings at these selected positions, the children strings produced after crossover will look as follows:

$$1\ 1\ 0\ 1\ 1\ 0\ 0\ 1\ 1\ 0\ 0\ 1\ 1\ 1\ 1\ 0$$

$$0\ 0\ 1\ 0\ 0\ 1\ 1\ 0\ 0\ 1\ 1\ 0\ 0\ 0\ 0\ 1$$

6.2.1.5 Mutation

In biology, mutation refers to a certain change in parameters on the gene level. In GA, the purpose of mutation operation is to create a solution in the vicinity of the current solution, thereby bringing a local change over the current solution. Thus, if a solution gets stuck at the local minimum, mutation may help it to come out of this situation and consequently, it may jump to the global basin. Hence, mutation helps to exploit the search space by optimizing within a promising area.

In mutation, 1 is converted into 0 and vice versa. To implement a bit-wise mutation scheme, we toss a coin with a probability of mutation (p_m) for every bit of a crossed-over pool. The value of p_m is generally kept low because with a higher value of p_m the genetic evolution degenerates into a random local search. If the outcome of the toss in a particular bit position is true, then that bit will be mutated, i.e., 1 will be converted into 0 and vice versa. For this purpose, a random number lying between 0 and 1 is generated at every bit position, and if it is found to be smaller or equal to p_m, the outcome is considered as true. To explain the principle of mutation, let us consider that prior to the mutation two strings are looking as follows:

$$1\ 1\ 1\ \ 1\ 1\ 1\ \boxed{1}\ 1\ 1\ 1\ 1\ 1\ 1\ 1\ 1$$

$$0\ 0\ \boxed{0}\ 0\ 0\ 0\ 0\ 0\ 0\ 0\ 0\ \boxed{0}\ 0\ 0\ 0\ 0$$

While performing the toss in every bit position, suppose that the outcomes are true in 7th position for the first string; then 3rd and 12th positions for the second string (marked by boxes). After mutation, the two strings will be looking as follows:

$$1\ 1\ 1\ 1\ 1\ 1\ 0\ 1\ 1\ 1\ 1\ 1\ 1\ 1\ 1$$

$$0\ 0\ 1\ 0\ 0\ 0\ 0\ 0\ 0\ 0\ 1\ 0\ 0\ 0\ 0$$

Because of crossover children acquire some common properties of their parents, but mutation introduces some new property for which children are little bit different than their parents. Crossover leads to a big jump to an area somewhere in between two parent solutions, whereas mutation causes random small diversion near a solution. After some repeated crossover without mutation, same solutions may repeat again and again. Thus, crossover alone may not be sufficient to reach the optimum solution. The role of mutation is explained with the help of Fig. 6.5, which illustrates the plot of a bimodal function having one local minima in the local basin and one global minima in the global basin. Let us assume that all randomly generated initial solutions fall on the local basin coincidentally. For example, consider the following population of initial solution:

$$\begin{array}{cccccccc} 0 & 0 & 1 & 0 & 1 & 1 & 0 & 1 \\ 0 & 1 & 1 & 1 & 0 & 0 & 1 & 0 \\ 0 & 0 & 1 & 0 & 1 & 1 & 1 & 0 \\ & & & \vdots & & & & \\ 0 & 1 & 0 & 1 & 0 & 0 & 0 & 1 \end{array}$$

It is noticed that all strings of the initial population have a 0 in their leftmost bit position. If the global optimum solution requires 1 in the leftmost bit position, then crossover operator alone cannot able to create 1 in that position. In this situation, mutation operator can only turn 0 into 1 and thereby creates the possibility of pushing a string from the local basin into the global basin as depicted in Fig. 6.5. Hence, mutation helps to achieve global optimum solution by maintaining small diversity in the population.

6.2.2   The flow chart of GA

The flow chart of GA is shown in Fig. 6.6. There are six steps as follows:

- Step 1: Initialize population size, maximum number of generation (gen_{max}), crossover probability (p_c), mutation probability (p_m), lower and upper

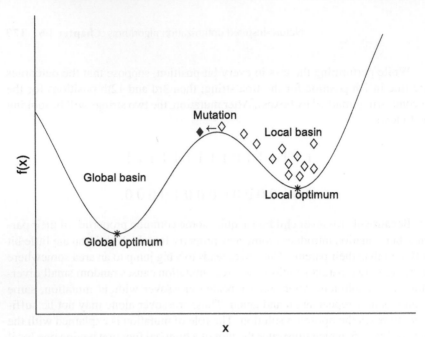

FIG. 6.5 Bimodal function with one local minimum and one global minimum.

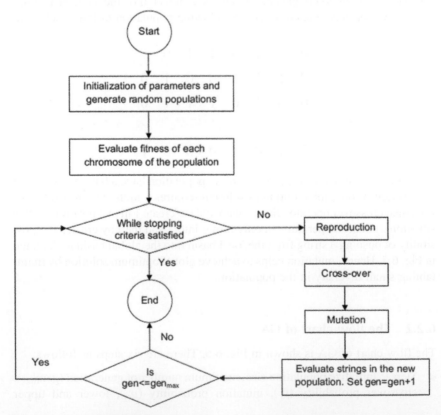

FIG. 6.6 Flow chart of GA.

bounds of variables, and bit size for each variable. Generate a random population of binary strings. Set generation (gen)$=0$.

- Step 2: Evaluate the fitness of each string in the population.
- Step 3: If gen $>$ gen$_{max}$ or other stopping criteria are satisfied, stop. Otherwise, carry out reproduction operation.
- Step 4: Execute crossover operation.
- Step 5: Perform mutation operation.
- Step 6: Evaluate strings in the new population, set gen $=$ gen $+1$, and go to step 3.

6.2.3 Working principle of GA in step by step

6.2.3.1 Unconstrained optimization problem

Let us consider the following unconstrained optimization problem with Himmelblau's function to explain the working principle of the GA:

$$\text{Minimize } f(x_1, x_2) = \left(x_1^2 + x_2 - 11\right)^2 + \left(x_1 + x_2^2 - 7\right)^2; \text{ for } 0 \leq x_1, x_2 \leq 5$$

(6.2)

where x_1 and x_2 are the real variables. As GA is designed for maximization, the minimization of the above function has been modified as

$$\text{Maximize } F(x_1, x_2) = \frac{1}{1 + f(x_1, x_2)}; \text{ for } 0 \leq x_1, x_2 \leq 5 \quad [\because f(x_1, x_2) \neq -1]$$

Here we denote $f(x_1, x_2)$ and $F(x_1, x_2)$ as the objective function and fitness function, respectively. Contour plot of the objective function $f(x_1, x_2)$ is displayed in Fig. 6.7. It can be shown that, the minimum value of the function $f(x_1, x_2)$ is 0, which is obtained at $x_1 = 3$ and $x_2 = 2$. In Fig. 6.7, the optimum solution is marked by an asterisk (*).

The following steps are involved to solve the above optimization problem using GA:

Step 1: Generation of initial population

An initial population of solutions of size N is selected at random. The solutions are expressed in binary coding for representing the variables x_1 and x_2. We have assigned 5 bits to represent each variable, hence each string of the population is 10 bits long. The population size (N) may become 100, 200, ..., 1000; depending on the complexity of the problem. However, for the sake of

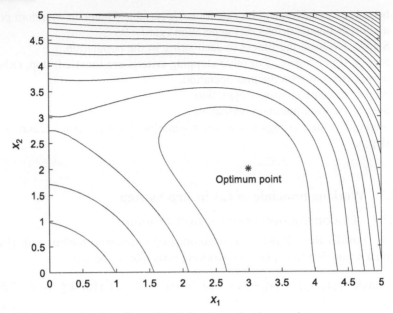

FIG. 6.7 Contour plot of the Himmelblau's function and optimum point.

simplicity, let us suppose that the initial population of only 8 strings is created at random shown as follows:

String 1: 1 0 0 0 0 1 0 1 0 0

String 2: 1 0 0 1 1 1 0 0 0 0

String 3: 0 0 1 1 1 1 0 1 0 0

String 4: 0 0 1 0 1 1 0 1 0 1

String 5: 1 0 0 1 0 0 1 0 1 0

String 6: 1 1 0 0 1 1 1 0 0 0

String 7: 0 1 0 0 1 0 0 1 1 1

String 8: 0 1 1 0 1 0 1 0 0 1

Each of the above strings is comprised of two substrings each of which is 5 bits long.

Step 2: Fitness evaluation

To determine the fitness value of each solution (that is, string), the real values of the variables x_1 and x_2 are to be calculated first. Knowing the minimum and maximum limits of a variable (say x_1) and decoded value of the binary substring assigned to represent it, the real value of variable x_1 can be determined using the linear mapping rule given in Eq. (6.1). For example, in the case of string number 1, the first substring (10000) is decoded as $0 \times 2^0 + 0 \times 2^1 + 0 \times 2^2 + 0 \times 2^3 + 1 \times 2^4 = 16$. Thus, the corresponding real value of variable x_1 is equal to $0 + \frac{(5-0) \times 16}{2^5 - 1}$ or 2.5806. Similarly, the decoded value of the second substring (10100) is worked out to be 20. Hence, the corresponding real value of variable x_2 is equal to $0 + \frac{(5-0) \times 20}{2^5 - 1}$ or 3.2258. Therefore, the first string corresponds to the solution $x_1 = 2.5806$ and $x_2 = 3.2258$. Consequently, the values of objective function $f(x_1, x_2)$ and fitness function $F(x_1, x_2)$ for the first string are evaluated as 37.0799 and 0.0263, respectively. The same procedure is repeated to evaluate the fitness values of the other strings of the population. Table 6.1 illustrates the decoded values, real values of variables, objective function values, and fitness function values of all 8 strings in the initial population. From the last column of Table 6.1, the average and maximum fitness values are estimated as 0.056 and 0.2407, respectively. Therefore, the ratio (r) between the average fitness and maximum fitness of the initial population becomes 0.2327. It appears from Table 6.1 that string number 5 is the best individual in the initial population and the corresponding values of x_1 and x_2 are 2.9032 and 1.6129, respectively. In Fig. 6.8, the initial population is marked with the empty diamonds on the contour plot of the objective function, whereas the best individual (string 5) is marked with a filled diamond.

Step 3: Reproduction

Here, we have used the roulette-wheel selection method for the purpose of parent selection. Table 6.2 depicts the working principle of roulette-wheel

TABLE 6.1 Fitness evaluation.

String no.	Substring 1	Substring 2	D_1	D_2	x_1	x_2	$f(x_1, x_2)$	$F(x_1, x_2)$
1	1 0 0 0 0	1 0 1 0 0	16	20	2.5806	3.2258	37.0799	0.0263
2	1 0 0 1 1	1 0 0 0 0	19	16	3.0645	2.5806	8.3661	0.1068
3	0 0 1 1 1	1 0 1 0 0	7	20	1.1290	3.2258	62.8082	0.0157
4	0 0 1 0 1	1 0 1 0 1	5	21	0.8065	3.3871	76.3435	0.0129
5	1 0 0 1 0	0 1 0 1 0	18	10	2.9032	1.6129	3.1545	0.2407
6	1 1 0 0 1	1 1 0 0 0	25	24	4.0323	3.8710	227.7581	0.0044
7	0 1 0 0 1	0 0 1 1 1	9	7	1.4516	1.1290	78.5407	0.0126
8	0 1 1 0 1	0 1 0 0 1	13	9	2.0968	1.4516	34.3602	0.0283

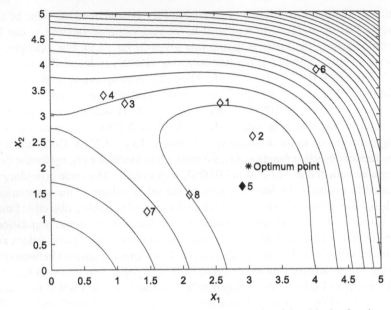

FIG. 6.8 Scattering of the initial population on the contour plot of the objective function.

selection. It begins with the computation of probability of selection (P) of each string on the basis of its fitness value. The value of P for each string is obtained as $F(x_1,x_2)/\sum F(x_1,x_2)$. Subsequently, the cumulative probability (Q) of each string is computed. In the next step, random numbers between 0 and 1 are created for spotting the particular string to be copied in the mating pool. Table 6.2 shows that the first random number is 0.5201, which lies between the cumulative probabilities 0.3611 and 0.8990, therefore, the 5th string is getting copied in the mating pool. In a similar manner, the other strings are copied in the mating pool. The last column of Table 6.2 shows the actual count of each string in the mating pool.

It is noted that mating pool contains multiple copies of a good string with a higher probability of selection. For example, string number 5 has been copied 5 times in the mating pool. The inferior strings with lower selection probabilities such as 1, 3, 4, and 6 are not included in the mating pool. As the selection of strings is made probabilistically, not all selected strings are better than all rejected strings. For example, the 7th string with a low selection probability of 0.0281 is copied but the 1st string with relatively more selection probability of 0.0587 has not been copied in the mating pool. The mating pool after reproduction from the initial population are shown as follows:

TABLE 6.2 Roulette wheel selection.

String no.	Substring 1	Substring 2	$F(x_1, x_2)$	P	Q	Random number	Selected string	Count in mating pool
1	1 0 0 0 0	1 0 1 0 0	0.0263	0.0587	0.0587	0.5201	5	0
2	1 0 0 1 1	1 0 0 0 0	0.1068	0.2386	0.2972	0.2518	2	2
3	0 0 1 1 1	1 0 1 0 0	0.0157	0.0350	0.3322	0.8372	5	0
4	0 0 1 0 1	1 0 1 0 1	0.0129	0.0289	0.3611	0.5263	5	0
5	1 0 0 1 0	0 1 0 1 0	0.2407	0.5378	0.8990	0.4358	5	5
6	1 1 0 0 1	1 1 0 0 0	0.0044	0.0098	0.9087	0.5850	5	0
7	0 1 0 0 1	0 0 1 1 1	0.0126	0.0281	0.9368	0.9098	7	1
8	0 1 1 0 1	0 1 0 0 1	0.0283	0.0632	1.0000	0.1098	2	0

String 5:	1	0	0	1	0	0	1	0	1	0
String 2:	1	0	0	1	1	1	0	0	0	0
String 5:	1	0	0	1	0	0	1	0	1	0
String 5:	1	0	0	1	0	0	1	0	1	0
String 5:	1	0	0	1	0	0	1	0	1	0
String 5:	1	0	0	1	0	0	1	0	1	0
String 7:	0	1	0	0	1	0	0	1	1	1
String 2:	1	0	0	1	1	1	0	0	0	0

Step 4: Crossover

The strings in the mating pools are subjected to the crossover operation. In this operation, an exchange of properties between the pair of parent strings takes place from which a pair of offspring strings are produced. In our example, 4 mating pairs are formed from a population of 8 strings. It is noted that not all 4 mating pairs are subjected to crossover. This is because we would like to retain some good solutions. For example, in a population of size 8, with $p_c = 0.7$, the approximate number of mating pairs that are subjected to crossover is $0.7 \times 8/2 \approx 3$. Table 6.3 shows that 3rd pair has not been subjected to crossover and its strings are simply copied in the intermediate population.

Uniform crossover has been used in this example. Table 6.3 depicts the working principle of uniform crossover. Let us consider the case of 1st mating pair as depicted in Table 6.3 where parent strings 5 and 2 are participated in crossover operation. While performing the tosses at each bit position, the outcomes turn out to be heads at position numbers 2, 3, 7, 9, and 10. Hence, swapping of the bits between the two parent strings 5 and 2 has taken place in these selected positions, which leads to the formation of two offspring strings. In a particular bit position of the two parent strings, if both are either 1 s or 0 s, the swapping of bits does not make any difference. Thus, in bit position numbers 2, 3, and 10 there is no effect after swapping. Once the crossover operation for the first mating pair is over, the second mating pair is selected for the same. The rest of the mating pairs also follow the similar process as depicted in Table 6.3. It is noted that for the 3rd pair, as the toss turns out to be tail, no crossover operation has taken place and the strings remain unchanged.

Step 5: Mutation

In this example, we have considered probability of mutation (p_m) as 0.001. In our example, while performing the toss at every bit of the crossed-over pool, the outcome turns out to be true only in the three positions that are marked by the box in Table 6.4. The mutation flipped the value in these bit positions to form the mutated pool, which is shown in the second column of Table 6.4.

TABLE 6.3 Crossover.

Pair no.	Whether crossover occurs?	Uniform crossover										
1	Yes	Parent strings	1	0	0	1	0	0	1	0	1	0
			1	0	0	1	1	1	0	0	0	0
		Toss results	T	H	H	T	T	T	H	T	H	H
		Offspring strings	1	0	0	1	0	0	0	0	0	0
			1	0	0	1	1	1	1	0	1	0
2	Yes	Parent strings	1	0	0	1	0	0	1	0	1	0
			1	0	0	1	0	0	1	0	1	0
		Toss results	H	T	T	H	H	T	H	T	T	T
		Offspring strings	1	0	0	1	0	0	1	0	1	0
			1	0	0	1	0	0	1	0	1	0
3	No	Strings remain unchanged	1	0	0	1	0	0	1	0	1	0
			1	0	0	1	0	0	1	0	1	0
4	Yes	Parent strings	0	1	0	0	1	0	0	1	1	1
			1	0	0	1	1	1	0	0	0	0
		Toss results	T	H	H	T	T	H	H	H	T	H
		Offspring strings	0	0	0	0	1	1	0	0	1	0
			1	1	0	1	1	0	0	1	0	1

Step 6: Fitness evaluation of the mutated pool

This step is closely similar to step 2. After estimating the decoded and real values of the substrings of the new population, i.e., the mutated pool, the values of the objective function as well as fitness function are evaluated as given in Table 6.5. With this step 1st generation of GA ends.

After the first generation, average and maximum values of fitness function become 0.1614 and 0.4757, respectively, which gives a ratio (r) of 0.3393. Therefore, the maximum value of fitness function increases from 0.2407 to 0.4757 after 1st generation. Further, the average value of fitness function has improved from 0.056 to 0.1614 and the value of r has increased from 0.2327

TABLE 6.4 Mutation.

Crossed-over pool								Mutated pool							
1	0	[0]	1	0	0	0	0	0	0	0	0	1	0	0	0
1	0	0	1	1	1	1	0	1	1	0	1	0	1	1	0
1	0	0	0	0	0	1	0	1	0	0	1	0	0	0	0
1	0	0	1	0	0	1	0	1	1	0	1	0	0	0	0
1	0	0	1	0	0	1	0	1	1	[0]	1	0	0	0	0
1	0	0	1	0	0	1	0	1	1	0	1	0	0	0	0
0	0	0	0	1	1	0	0	1	1	[0]	0	0	1	1	0
1	1	0	1	1	0	0	1	0	0	1	0	0	1	1	1

TABLE 6.5 Fitness evaluation.

String no.	Substring 1	Substring 2	D_1	D_2	x_1	x_2	$f(x_1, x_2)$	$F(x_1, x_2)$
1	1 0 1 1 0	0 0 0 0 0	22	0	3.5484	0	14.4451	0.0647
2	1 0 0 1 1	1 1 0 1 0	19	26	3.0645	4.1935	193.0137	0.0052
3	1 0 0 1 0	0 1 0 1 0	18	10	2.9032	1.6129	3.1545	0.2407
4	1 0 0 1 0	0 1 0 1 0	18	10	2.9032	1.6129	3.1545	0.2407
5	1 0 0 1 0	0 1 1 1 0	18	14	2.9032	2.2581	1.1023	0.4757
6	1 0 0 1 0	0 1 0 1 0	18	10	2.9032	1.6129	3.1545	0.2407
7	0 0 0 0 1	1 0 1 1 0	1	22	0.1613	3.5484	88.2289	0.0112
8	1 1 0 1 1	0 0 1 0 1	27	5	4.3548	0.8065	80.9109	0.0122

to 0.3393 after 1st generation. Table 6.5 shows that string number 5 is the best individual in the new population and the corresponding values of x_1 and x_2 are 2.9032 and 2.2581, respectively. Fig. 6.9 depicts the new population after 1st generation (marked with the empty diamonds) and the best individual (marked with the filled diamond).

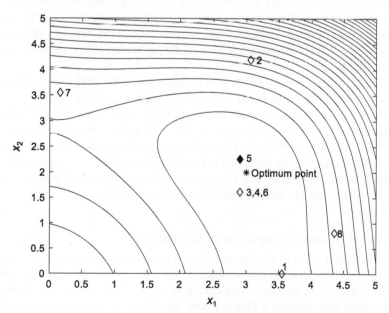

FIG. 6.9 Scattering of the new population on the contour plot of the objective function.

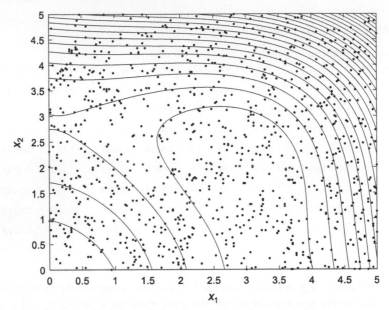

FIG. 6.10 Scattering of initial population of size 1000 on the contour plot.

Once the first generation is completed we proceed to the next generation, which resumes from step 3. It is expected that in each generation the values of average fitness and r are improved. Steps 3–6 are repeated until the stopping criterion is satisfied. The stopping criterion may be set either based on the maximum number of generations or a desired value of r. If the desired value of r is chosen as 0.99, then on an average 99% of the population converges to the maximum fitness value.

The MATLAB code of GA with a population size of 1000 was executed to solve the optimization problem where 8 bits were assigned to represent each variable. Fig. 6.10 shows the spread of initial population of size 1000 in the search space of two variables x_1 and x_2. After 575 generations the stopping criterion ($r > 0.99$) was satisfied. All 1000 points in the population after meeting the stopping criterion are displayed in Fig. 6.11, from which it is manifested that majority of the population have either reached the optimum solution ($x_1 = 3$ and $x_2 = 2$) or they are clustered around the optimum solution.

6.2.3.2 Constrained optimization problem

If the optimum point of the unconstrained optimization violates the constraint function, then the optimum point of the constrained optimization will be different than that of the unconstrained optimization. Let us consider the following constrained optimization problem where the objective function is as same as the unconstrained optimization problem.

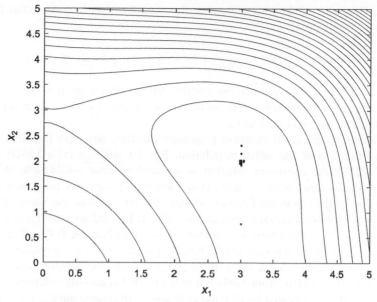

FIG. 6.11 Scattering of population on the contour plot after meeting the stopping criterion.

$$\text{Minimize } f(x_1, x_2) = \left(x_1^2 + x_2 - 11\right)^2 + \left(x_1 + x_2^2 - 7\right)^2; \text{ for } 0 \leq x_1, x_2 \leq 5$$
$$\text{Subject to } (x_1 - 5)^2 + x_2^2 - 25 \geq 0$$

$$(6.3)$$

In the above-constrained optimization problem, the original optimum point $x_1 = 3$ and $x_2 = 2$ provides a value of the constraint function less than zero, thus it becomes an infeasible point. To handle the constraint function, a bracket operator penalty term has been used, which is defined as follows:

$$\Omega = R\langle g(x) \rangle^2 \qquad (6.4)$$

where $\langle \alpha \rangle = \alpha$, when α is negative; otherwise $\langle \alpha \rangle = 0$. The inclusion of bracket penalty operator term leads to the minimization of the following penalty function (pf) for solving the constrained optimization problem:

$$pf = f(x_1, x_2) + R\langle g(x) \rangle^2 \qquad (6.5)$$

Here, we choose a value of R as 100. Therefore, in this example, the optimization problem becomes

$$\text{Minimize } pf = \left(x_1^2 + x_2 - 11\right)^2 + \left(x_1 + x_2^2 - 7\right)^2$$
$$+ 100\left\langle (x_1 - 5)^2 + x_2^2 - 25 \right\rangle^2; \text{ for } 0$$
$$\leq x_1, x_2 \leq 5$$

As GA is designed for maximization, the minimization of the above function has been modified as

$$\text{Maximize } F(x_1, x_2) = \frac{1}{1 + pf}; \text{ for } 0 \le x_1, x_2 \le 5 \quad [\because pf \ne -1]$$

To explain the above constrained optimization problem in step-by-step, we have employed the same initial population as in the unconstrained optimization discussed in the previous section.

Table 6.6 depicts the values of constraint function, objective function, and fitness function of the initial population. For 1st solution ($x_1 = 2.5806$ and $x_2 = 3.2258$), the constraint function is violated because $-8.741 < 0$. When the constraint function is violated the bracket penalty operator term increases the value of the penalized function. Hence, for 1st solution, the value of the penalized function becomes large that is 7677.4. For 3rd solution ($x_1 = 1.129$ and $x_2 = 3.2258$), the constraint function is satisfied because $0.39 > 0$. When the constraint function is satisfied the bracket penalty operator term equals zero. Hence, for 3rd solution, the value of the penalized function becomes small that is 62.808. It is evident from Table 6.6 that except 3rd and 4th solutions, constraint function is violated for all other solutions. It thus turns out that the values of fitness function as well as probability of selection are significantly higher for 3rd and 4th solutions than the other solutions. Consequently, there will be high possibility of copying these two solutions during the formation of mating pool. Table 6.7 shows that strings 3 and 4 have been selected 5 and 3 times, respectively, using the roulette wheel selection. Next, we proceed to the crossover, mutation, and fitness evaluation sequentially as discussed in the previous section. With this the first generation of GA completes and the next generation resumes from step 3. Steps 3 to 6 will be repeated until the stopping criteria are reached.

The MATLAB code of GA with a population size of 1000 each with 16 bits was executed to solve this constrained optimization problem. After 668 generations the stopping criterion ($r > 0.99$) was satisfied and the optimum solution was found to be $x_1 = 0.90196$ and $x_2 = 2.8627$. Fig. 6.12 shows the feasible region and optimum point.

6.2.4 Application of GA in fabric engineering

The term "fabric engineering" refers to the manufacturing of fabrics with desired quality at a low cost of production by optimizing fabric parameters such as yarn count, crimp, and thread spacing both in warp and weft directions. The intricate relationship of fabric parameters with its physical and mechanical properties makes it too complex to solve this optimization problem by means of traditional technique. The excellent searching abilities of GA facilitate to select the best combinations of parameters to obtain fabrics with the desired quality with low manufacturing cost. It is necessary to have an appropriate areal

TABLE 6.6 Fitness evaluation.

String no.	Substring 1	Substring 2	x_1	x_2	Constraint function	pf	$F(x_1, x_2)$
1	1 0 0 0 0	1 0 1 0 0	2.5806	3.2258	−8.741	7677.4	0.00013024
2	1 0 0 1 1	1 0 0 0 0	3.0645	2.5806	−14.594	21,307	0.0000469
3	0 0 1 1 1	1 0 1 0 0	1.1290	3.2258	0.39	62.808	0.0156720
4	0 0 1 0 1	1 0 1 0 1	0.8065	3.3871	4.058	76.343	0.0129290
5	1 0 0 1 0	0 1 0 1 0	2.9032	1.6129	−18.002	32,411	0.0000309
6	1 1 0 0 1	1 1 0 0 0	4.0323	3.8710	−9.079	8470.7	0.0001180
7	0 1 0 0 1	0 0 1 1 1	1.4516	1.1290	−11.134	12,476	0.0000802
8	0 1 1 0 1	0 1 0 0 1	2.0968	1.4516	−14.464	20,955	0.0000477

TABLE 6.7 Roulette wheel selection.

String no.	Substring 1	Substring 2	$F(x_1, x_2)$	P	Q	Random number	Selected string	Count in mating pool
1	1 0 0 0 0	1 0 1 0 0	0.00013024	0.004482	0.004482	0.5172	3	0
2	1 0 0 1 1	1 0 0 0 0	0.0000469	0.001615	0.006098	0.7193	4	0
3	0 0 1 1 1	1 0 1 0 0	0.0156720	0.53939	0.54548	0.33218	3	5
4	0 0 1 0 1	1 0 1 0 1	0.0129290	0.44499	0.99047	0.1367	3	3
5	1 0 0 1 0	0 1 0 1 0	0.0000309	0.001062	0.99154	0.95978	4	0
6	1 1 0 0 1	1 1 0 0 0	0.0001180	0.004063	0.9956	0.25999	3	0
7	0 1 0 0 1	0 0 1 1 1	0.0000802	0.002759	0.99836	0.70263	4	0
8	0 1 1 0 1	0 1 0 0 1	0.0000477	0.001642	1	0.055287	3	0

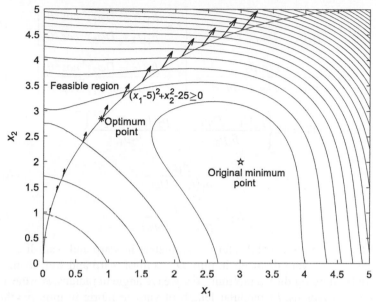

FIG. 6.12 Feasible region and optimum point for constrained optimization problem.

density or GSM of fabric for a good quality as well as economic production. Fabric with very low GSM is too sparse to qualify for the desirable quality, whereas higher GSM warrants higher manufacturing costs entailing more yarn consumption. Nevertheless, sometimes fabrics manufactured with fine yarns are costlier than the heavy fabrics due to higher price of fine yarns. Thus, it is necessary to maintain a particular range of yarn count while optimizing fabric cost by minimizing GSM. Three different ranges of counts for warp and weft yarns, i.e., fine (10–24 tex), medium (30–50 tex), and coarse (60–100 tex) made from 100% cotton fibers are considered for the production of light, medium and heavy weight fabrics, respectively. The optimization problem is formulated to minimize the GSM (W) of these fabrics such that different physical and mechanical parameters both in warp and weft directions viz., cover (K), tensile modulus (E), shear rigidity (G) and flexural rigidity (B) are retained within the desired levels and the inter yarn forces in the fabric are balanced in its relaxed state. The rigid thread model (the elastica model) of fabric with plain weave has been considered in this study. The expressions of W, K, E, G, and B are given in Eqs. (6.6)–(6.12) (Leaf et al., 1993; Leaf & Kandil, 1980; Leaf & Sheta, 1984):

$$W = \frac{T_1(1 + c_1)}{p_1} + \frac{T_2(1 + c_2)}{p_2} \tag{6.6}$$

$$K = \frac{d_1}{p_1} + \frac{d_2}{p_2} - \frac{d_1 d_2}{p_1 p_2} \tag{6.7}$$

$$E_1 = \frac{12 p_2 \beta_1}{p_1 (l_1 - D\varphi_1)^3 \sin^2\theta_1} \times \left[1 + \frac{\beta_2 (l_1 - D\varphi_1)^3 \cos^2\theta_1}{\beta_1 (l_2 - D\varphi_2)^3 \cos^2\theta_2} \right] \tag{6.8}$$

$$E_2 = \frac{12 p_1 \beta_2}{p_2 (l_2 - D\varphi_2)^3 \sin^2\theta_2} \times \left[1 + \frac{\beta_1 (l_2 - D\varphi_2)^3 \cos^2\theta_2}{\beta_2 (l_1 - D\varphi_1)^3 \cos^2\theta_1} \right] \tag{6.9}$$

$$G = 12 \left\{ \frac{p_1 (l_1 - D\varphi_1)^3}{\beta_1 p_2} + \frac{p_2 (l_2 - D\varphi_2)^3}{\beta_2 p_1} \right\}^{-1} \tag{6.10}$$

$$B_1 = \frac{\beta_1 p_2}{p_1 (l_1 - D\varphi_1)} \tag{6.11}$$

$$B_2 = \frac{\beta_2 p_1}{p_2 (l_2 - D\varphi_2)} \tag{6.12}$$

where the subscripts 1 and 2 refer to parameters in warp and weft directions, respectively, d = yarn diameter in mm, D = sum of warp and weft diameters, β = yarn bending rigidity in mN mm^2, θ = weave angle in radians, ϕ = the angle of contact in radians, l = modular length of yarn in fabric in mm, p = thread spacing in mm, c = yarn crimp in the fabric in fraction, T = yarn count in tex. The following empirical relationship between θ and c provides a reasonable agreement between theoretical and experimental values (Leaf, 2003):

$$\theta = 1.88 c^{\frac{1}{2}} \tag{6.13}$$

Leaf (2003) established the relationship between angle of contact (ϕ) and weave angle (θ) for the rigid thread model as follows:

$$\varphi = \sin^{-1}\left(2 \sin\theta - \frac{p}{D} \right) \tag{6.14}$$

The yarn diameter (d in mm) is related to its tex (T) by the following expression (Peirce, 1937):

$$d = 0.0357 \sqrt{v_y T} \tag{6.15}$$

where v_y = yarn-specific volume in cc/g. For cotton yarn the value of v_y is 1.1 cc/g. The expression of yarn bending rigidity is given by (Postle et al., 1988):

$$\beta = \beta_f t_f^2 n \tag{6.16}$$

where β_f = specific fiber rigidity in mN mm^2/tex^2, t_f = fiber tex and n = number of fiber in the yarn cross section. For cotton fiber the value of β_f is 0.53 mN mm^2/tex^2 (Morton & Hearle, 1993).

At relaxed state of a fabric, the internal forces acting vertically on the warp and weft threads due to the resistance of bending are balanced if the following equation is satisfied (Peirce, 1937):

$$\frac{\beta_1 \sin \theta_1}{p_2^2} = \frac{\beta_2 \sin \theta_2}{p_1^2} \tag{6.17}$$

Eq. (6.17) has predominant influence on the balance of crimp and therefore known as crimp balance equation.

The following optimization problem is developed to select the fabric governing parameters such as yarn counts (T_1 and T_2), crimps (c_1 and c_2), and thread spacing (p_1 and p_2) for the production of light-, medium-, and heavy-weight fabrics so as to manufacture them with minimum cost as well as requisite quality.

$$\left. \begin{array}{c} \text{Minimize } W \\ \text{subject to the inquality constraints :} \\ K^U \geq K \geq K^L \\ E_1{}^U \geq E_1 \geq E_1{}^L \\ E_2{}^U \geq E_2 \geq E_2{}^L \\ G^U \geq G \geq G^L \\ B_1{}^U \geq B_1 \geq B_1{}^L \\ B_2{}^U \geq B_2 \geq B_2{}^L \\ \text{and equality constraint :} \\ \dfrac{\beta_1 \sin \theta_1}{p_2^2} = \dfrac{\beta_2 \sin \theta_2}{p_1^2} \end{array} \right\} \tag{6.18}$$

where the superscripts L and U refer to the values of lower and upper bounds, respectively. A penalty function method is used to handle the constraints. A bracket penalty operator is used to handle the inequality constraints whereas the equality constraint was handled by a parabolic penalty operator (Deb, 2005). Accordingly, for minimizing the objective function $f(x)$ subject to equality constraints $h(x)=0$ and inequality constraints $g(x)\leq0$, the penalty function is formulated as

$$\underbrace{P(x)}_{\text{Min}} = f(x) + \mu\left[\{h(x)\}^2 + \{\max(0, g(x)\}^2\right] \tag{6.19}$$

where μ is the penalty parameter. The search space for the governing parameters determining the physical and mechanical properties of different fabrics is given in Table 6.8. Table 6.9 shows the lower and upper bounds of inequality constraints for three types of fabrics. The optimization problem of Eq. (6.18) has been solved using GA with MATLAB coding.

At the outset, a binary-coded population size of 2000 is randomly generated. Each individual of the population represents 6 weaving parameters such as yarn counts, crimps, and thread spacing for warp and weft directions. For each parameter, 8 bits are chosen, which makes total string length of an individual to 48. The binary-coded parameters are then converted into real value by a linear

TABLE 6.8 Search space for the fabric governing parameters.

Fabric parameters	Light weight fabric (using fine yarns)		Medium weight fabric (using medium count yarns)		Heavy weight fabric (using coarse yarns)	
	Lower bound	*Upper bound*	*Lower bound*	*Upper bound*	*Lower bound*	*Upper bound*
T_1, tex	10	24	30	50	60	100
T_2, tex	10	24	30	50	60	100
c_1, fraction	0.03	0.15	0.03	0.15	0.03	0.15
c_2, fraction	0.03	0.15	0.03	0.15	0.03	0.15
p_1, mm	0.20	0.40	0.45	0.80	0.50	0.85
p_2, mm	0.20	0.40	0.45	0.80	0.50	0.85

TABLE 6.9 Boundary of the constraints for different types of fabrics.

Fabric properties	Light weight fabrics		Medium weight fabrics		Heavy weight fabrics	
	Lower bound	*Upper bound*	*Lower bound*	*Upper Bound*	*Lower bound*	*Upper bound*
K	0.6	0.72	0.6	0.72	0.6	0.72
E_1, mN/mm	1000	2500	2000	3500	3500	5500
E_2, mN/mm	1000	2500	2000	3500	3500	5500
G, mN/mm	50	150	100	250	200	350
B_1, mN mm	3	12	6	15	12	20
B_2, mN mm	3	12	6	15	12	20

mapping. Since GA is a maximizing procedure, the penalty function in Eq. (6.19) is modified to the following fitness function

$$F(X) = \frac{1}{1 + \{P(X)\}^2} \tag{6.20}$$

The above fitness function is then evaluated for whole population from which the ratio of the average fitness value to the maximum fitness value (r) is calculated. The initial population is then modified using different operators of GA, namely reproduction, crossover (p_c), and mutation (p_m). A new population is now formed from which the fitness is evaluated and r is recalculated. This completes the first generation of GA. The GA runs for generation after generation until it satisfies the termination criteria, which is either r attains a desired value or the number of generations reaches a maximum value. The desired value of r is chosen as 0.99 meaning 99% of the population converges to the optimum fitness value. The maximum number of generations was set to 1000. The optimum solution of GA is obtained with the values of 0.8 and 0.001 for p_c and p_m, respectively. The roulette wheel selection scheme is applied for reproduction operation to select the good individuals of weaving parameters from the population on the basis of their fitness information. The uniform crossover method is applied to form new individuals.

The optimum values of the fabric governing parameters for production of light-, medium-, and heavy-weight fabrics obtained by solving the optimization problem of Eq. (6.18) are summarized in Table 6.10. The physical and mechanical properties of fabrics are predicted from the obtained parameters of Table 6.10 using the Eqs. (6.6)–(6.12). Table 6.11 depicts the predicted values

TABLE 6.10 Optimum values of governing parameters.

Fabric type	T_1, tex	T_2, tex	c_1, fraction	c_2, fraction	p_1, mm	p_2, mm
Light weight	13	14	0.103	0.101	0.398	0.396
Medium weight	48	43	0.040	0.062	0.784	0.741
Heavy weight	80	77	0.094	0.110	0.816	0.802

TABLE 6.11 Obtained fabric properties.

Fabric type	K	E_1, mN/ mm	E_2, mN/ mm	G, mN/ mm	B_1, mN mm	B_2, mN mm	W, g/m^2
Light weight	0.57	1843	1909	149.2	3.0	3.2	74.9
Medium weight	0.55	3096	2113	105.7	6.2	6.0	125.3
Heavy weight	0.65	4236	3582	309.8	12.6	12.6	213.8

of the physical and mechanical properties of three types of fabrics. It is evident from Table 6.11 that the predicted properties of all three fabrics are lying within the desirable limits. The optimized fabric parameters given in Table 6.10 also satisfy the Eq. (6.17). This ensures the balance of inter yarn forces in the fabric, which in turn effects the balance of warp and weft crimp. The obtained values of fabric GSM are also shown in Table 6.11.

6.3 Particle swarm optimization (PSO)

PSO is a population-based stochastic optimization technique inspired by the social behavior of bird flocking. There are many similarities between particle swarm optimization and genetic algorithms. Both the methods are initialized with a population of random solutions and the searches for optimal solutions are proceeded by updating the population in the next iterations. However, unlike genetic algorithms, PSO has no evolution operators such as crossover and mutation. PSO simulates the choreography of a bird swarm that flies through the problem space where each bird adjusts its traveling speed dynamically corresponding to the flying experiences of itself and its colleagues (Kennedy and Eberhart 1995).

PSO is initialized by a population of random solutions and each potential solution is assigned a randomized velocity. The potential solutions, called particles, are then flown through the problem space. The fitness function is evaluated for each particle. Each particle keeps track of its coordinates in the problem space, which is associated with the best solution or fitness value achieved so far. This value is called personal best or *pbest*. Another best value that is tracked by the global version of the particle swarm optimizer is the overall best value and its location, obtained so far by any particle in the population. This value is termed global best or *gbest*. Suppose that the location and velocity of an ith particle of the swarm at a time step t are represented by N-dimensional vectors $X_i^{(t)} = (x_{i1}, x_{i2}, ..., x_{iN})$ and $V_i^{(t)} = (v_{i1}, v_{i2}, ..., v_{iN})$, respectively. If the best positions experienced so far by the ith particle and the whole swarm are denoted as $pbest_i^{(t)} = (p_{i1}, p_{i2}, ..., p_{iN})$ and $gbest_i^{(t)}$, respectively, then in the next time step $t+1$, the ith particle of the swarm updates its velocity as well as location as follows (Shi & Eberhart, 1998a):

$$V_i^{(t+1)} = wV_i^{(t)} + C_1 r_1 \left(pbest_i^{(t)} - X_i^{(t)} \right) + C_2 r_2 \left(gbest^{(t)} - X_i^{(t)} \right) \quad (6.21)$$

$$X_i^{(t+1)} = X_i^{(t)} + V_i^{(t+1)} \quad (6.22)$$

where $i = 1, 2, ..., S$ (size of the swarm), w is the inertia weight, C_1 and C_2 are the learning factors, r_1 and r_2 are the random numbers uniformly distributed in the range (0, 1). According to Eqs. (6.21) and (6.22), the next displacement of a particle depends upon three fundamental elements such as inertia effect (its own velocity), personal influence (its best performance), and social influence

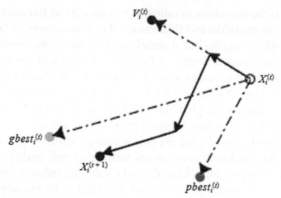

FIG. 6.13 Schematic movement of a particle.

(best performance of the group). In Eq. (6.21), the first term $wV_i^{(t)}$ is the inertia effect, responsible for keeping the particle moving in the same direction it was originally heading. The second term $C_1 r_1 \left(pbest_i^{(t)} - X_i^{(t)} \right)$ is the personal influence, accountable for the local search and it causes the particle to return to its best position it has experienced so far. The third term $C_2 r_2 \left(gbest^{(t)} - X_i^{(t)} \right)$ is the social influence, liable for global search by causing the particle to move to the best region the swarm has found so far. A schematic movement of a particle is depicted in Fig. 6.13 from which it is evident that the next position $X_i^{(t+1)}$ of a particle is determined by its current position $X_i^{(t)}$, its current velocity $V_i^{(t)}$, its own best performance achieved so far $pbest_i^{(t)}$, and the best performance received from its neighbors so far $gbest_i^{(t)}$.

It is essential to choose optimum values for the parameters for the best performance of PSO for different types of applications. So, the selections of the important parameters, such as inertia weight (w), learning factors (C_1, C_2), $pbest$ and $gbest$ have to be taken care of, which are discussed as follows.

The inertia weight w is utilized to adjust the influence of the previous velocity on the current velocity and to balance between global and local exploration abilities of the flying particle (Umapathy et al., 2010). A larger inertia weight implies stronger global exploration ability (i.e., searching new areas), which helps the particle to escape from local minima. A smaller inertia weight leads to stronger local exploration ability, which confines the particle searching within a local range near its present position and thus helps convergence. Hence, a proper selection of the inertia weight can provide a balance between global and local exploration abilities to reduce the average number of iterations. Shi and Eberhart (1998b) put forward the concept of linearly decreasing inertia weight for a better search as follows:

$$w = (w_{max} - w_{min}) \times \frac{iter_{max} - iter}{iter_{max}} + w_{min} \qquad (6.23)$$

where $iter_{max}$ is the maximum iteration, $iter$ is the current iteration number, and w_{max} and w_{min} are the initial and final values of w, respectively. A larger value of the inertia weight during the initial iterations promotes global exploration of the search space and a gradual decline of the inertia weight with the number of iterations promotes local exploration of the search space. Generally, it has been found that the values of 1.2 and 0.2 for w_{max} and w_{min}, respectively, can accelerate the convergence rate. Thus, an initial value of around 1.2 and, a gradual decline toward 0.2 can be considered a good choice for w.

The constants C_1 and C_2 are the two learning factors, which represent the weighting of the stochastic acceleration terms that pull each particle toward the *pbest* and *gbest* positions. Thus C_1 and C_2 are the balance factors between the effect of self-knowledge and social knowledge in moving the particle toward the target. The adjustments of these constants change the amount of tension in the system. Low values allow particles to roam far from target regions before being tugged back, while high values result in abrupt movements toward or past the target regions. The constants C_1 and C_2 are also termed as cognitive coefficient and social coefficient, respectively. The cognitive parameter represents the tendency of individuals to duplicate past behaviors that have proven successful, whereas the social parameter represents the tendency to follow the successes of others. Generally, C_1 and C_2 are set to 2.0, which will make the search cover all surrounding regions centered at *pbest* and *gbest*. Also, if the learning factors are identical, the same importance is given to local (nearby) searching and global (wide-ranging) searching, hence both parts contribute equally to the success of particle swarm searching.

6.3.1 The flow chart of particle swarm optimization

- Step 1: Initialization:
  Initialize swarm size, maximum number of iterations, number of variables, and lower and upper boundary of variables. Set the values of w, C_1, and C_2. Generate initial positions of the swarm X^0, which are uniformly distributed in the search space. Set $t=0$.
- Step 2: Fitness evaluation

Evaluate each particle's position according to the objective function.

- Step 3: Updating *pbest*

If a particle's current position is better than its previous best position, update it. Compare each particle's current position with its previous best position in terms of fitness value and if the current position of a particle is found to be better than its previous best position, assign the current position as *pbest*, else its previous best position is *pbest*.

- Step 4: Updating *gbest*

If the current population's overall best position is better than the population's overall previous best position, update it to *gbest*.

- Step 5: Updating the velocity of each particle

Update the velocity of each particle using Eq. (6.21).

- Step 6: Updating the position of each particle

Move each particle to its new position according to Eq. (6.22).

If the desired criterion is not met, set $t = t + 1$ and go to step 2; otherwise stop the process. When the process is stopped the optimum solution is given by *gbest*. A flowchart of the PSO algorithm is illustrated in Fig. 6.14.

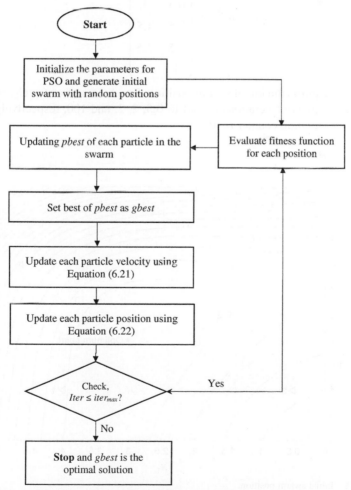

FIG. 6.14 Flowchart of particle swarm optimization algorithm.

6.3.2 Working principle of PSO in step-by-step

Let us consider the same unconstrained optimization problem of Section 6.2.3.1 to explain the working principle of the PSO as follows.

Step 1: Initialization

The typical value of swarm size S is around 1000; however, for the sake of simplicity, here we have considered $S=6$. Since there are two variables x_1 and x_2 in the optimization problem, it becomes a two-dimensional search space, which is bounded by $0 \leq x_1, x_2 \leq 5$. Suppose that within the given boundaries of search space, the swarm has the following initial positions:

$$X^{(0)} = \begin{bmatrix} 4 & 1 \\ 1.5 & 2.5 \\ 0.5 & 1 \\ 2.5 & 4.5 \\ 4.5 & 3.5 \\ 1 & 4 \end{bmatrix}$$

Fig. 6.15 shows the initial swarm position. The values of w, C_1, C_2, and the maximum number of iterations are set to 0.5, 2, 2, and 100, respectively. We also set the iteration counter $t=0$.

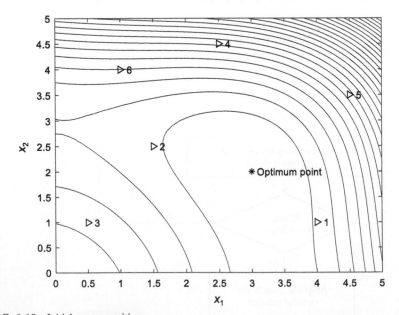

FIG. 6.15 Initial swarm position.

TABLE 6.12 Fitness values of swarm at initial positions.

Particle no.	$X^{(0)}$	$f(X^{(0)})$	$pbest^{(0)}$	$gbest^{(0)}$
1	[4 1]	40	[4 1]	[1.5 2.5]
2	[1.5 2.5]	39.63	[1.5 2.5]	[1.5 2.5]
3	[0.5 1]	125.31	[0.5 1]	[1.5 2.5]
4	[2.5 4.5]	248.13	[2.5 4.5]	[1.5 2.5]
5	[4.5 3.5]	257.63	[4.5 3.5]	[1.5 2.5]
6	[1 4]	136	[1 4]	[1.5 2.5]

Step 2: Fitness evaluation

The values of objective function $f(X^{(0)})$ for each particle in the swarm at its initial position $X^{(0)}$ are evaluated. Table 6.12 shows the objective function values of all 6 particles.

Step 3: Updating pbest

We assume that initial $pbest^{(0)}$ values are equal to $X^{(0)}$.

Step 4: Updating gbest

As the optimization problem is a minimization one, it is evident from Table 6.12 that particle number 2 at its initial position [1.5 2.5] gives the overall best value. Thus, the initial $gbest^{(0)}$ is [1.5 2.5].

Step 5: Updating the velocity of each particle

We set the initial velocity of swarm $V^{(0)}$ to be equal to $0.1 \times X^{(0)}$. Table 6.13 shows the initial velocity of the swarm. As the optimization problem has two variables, both the random numbers r_1 and r_2 will also have two dimensions for each particle. Table 6.13 shows the randomly generated values of r_1 and

TABLE 6.13 Initial and updated velocities of swarm.

Particle no.	$V^{(0)}$	r_1	r_2	$V^{(1)}$
1	[0.4 0.1]	[0.42 0.73]	[0.18 0.37]	[−0.7 1.16]
2	[0.15 0.25]	[0.84 0.57]	[0.73 0.18]	[0.08 0.13]
3	[0.05 0.1]	[0.96 0.92]	[0.27 0.22]	[0.57 0.71]
4	[0.25 0.45]	[0.37 0.64]	[0.09 0.18]	[−0.05 −0.49]
5	[0.45 0.35]	[0.05 0.35]	[0.72 0.66]	[−4.10 −1.15]
6	[0.1 0.4]	[0.38 0.02]	[0.63 0.91]	[0.68 −2.53]

r_2 for all 6 particles. The velocity of each particle is updated according to Eq. (6.21). For example, the updated velocity of the first particle is estimated as follows:

$$V_1{}^{(1)} = wV_1{}^{(0)} + C_1 r_1 \left(pbest_1{}^{(0)} - X_1{}^{(0)} \right) + C_2 r_2 \left(gbest^{(0)} - X_1{}^{(0)} \right)$$

$$= \begin{bmatrix} 0.5 \times 0.4 \\ 0.5 \times 0.1 \end{bmatrix}^T + \begin{bmatrix} 2 \times 0.42 \times (4 - 4) \\ 2 \times 0.73 \times (1 - 1) \end{bmatrix}^T + \begin{bmatrix} 2 \times 0.18 \times (1.5 - 4) \\ 2 \times 0.37 \times (2.5 - 1) \end{bmatrix}^T$$

$$= [-0.7 \ 1.16]$$

Similarly, other particles in the swarm update their velocities. The updated velocities of the whole swarm are given in Table 6.13.

Step 6: Updating the position of each particle

Each particle in the swarm will fly to its new position according to Eq. (6.22). For example, the new position of the first particle can be computed as follows:

$$X_1{}^{(1)} = X_1{}^{(0)} + V_1{}^{(1)}$$
$$= [4 \ 1] + [-0.7 \ 1.16]$$
$$= [3.3 \ 2.16]$$

Table 6.14 shows the new positions of all 6 particles. With this step, 1st iteration of particle swarm optimization is completed. The change of swarm position after 1st iteration is displayed in Fig. 6.16. We then proceed to the next iteration. Steps 2–6 will be repeated iteration after iteration until the maximum iteration is reached.

After 1st iteration, swarm occupy the new position $X^{(1)}$. The values of objective function $f(X^{(1)})$ in the newly occupied position of swarm are shown in Table 6.15. The fitness value of each particle in its new position is compared with the fitness value of its previous best position. In this example, we observed that all particles in the swarm show better fitness values in their new positions

TABLE 6.14 New position of swarm after 1st iteration.

Particle no.	$X^{(0)}$	$V^{(1)}$	$X^{(1)}$
1	[4 1]	[−0.70 1.16]	[3.3 2.16]
2	[1.5 2.5]	[0.08 0.13]	[1.57 2.63]
3	[0.5 1]	[0.57 0.71]	[1.06 1.71]
4	[2.5 4.5]	[−0.05 − 0.49]	[2.44 4]
5	[4.5 3.5]	[−4.10 − 1.15]	[0.4 2.35]
6	[1 4]	[0.68 − 2.53]	[1.68 1.47]

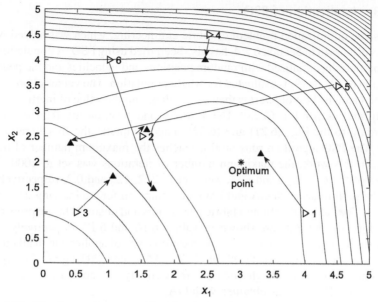

FIG. 6.16 The change of swarm position after 1st iteration.

TABLE 6.15 Fitness values, $pbest^{(1)}$ and $gbest^{(1)}$ of swarm after 1st iteration.

Particle no.	$X^{(1)}$	$f(X^{(1)})$	$pbest^{(1)}$	$gbest^{(1)}$
1	[3.3 2.16]	5.13	[3.3 2.16]	[3.3 2.16]
2	[1.57 2.63]	36.89	[1.57 2.63]	[3.3 2.16]
3	[1.06 1.71]	75.58	[1.06 1.71]	[3.3 2.16]
4	[2.44 4]	132.94	[2.44 4]	[3.3 2.16]
5	[0.4 2.35]	73.03	[0.4 2.35]	[3.3 2.16]
6	[1.68 1.47]	54.97	[1.68 1.47]	[3.3 2.16]

than their previous best positions. Hence, for all particles we assign their new positions as $pbest^{(1)}$. The updated values of $pbest^{(1)}$ after 1st iteration are put into Table 6.15. It is manifested from Table 6.15 that particle number 1 at its new position [3.3 2.16] gives the minimum value of fitness function among the swarm. Hence, after 1st iteration $gbest^{(1)}$ is assigned to [3.3 2.16].

The MATLAB® code of particle swarm optimization was executed with 1000 swarm and 100 iterations to solve the optimization problem of Eq. (6.2). The optimum solution was found to be $x_1 = 3$ and $x_2 = 2$ with the function value $= 0$.

6.3.3 Application of PSO

The same optimization problem of fabric engineering, which was solved with GA as discussed in Section 6.2.4, has also been tried using PSO. A population of 500 particles is initialized with random positions and velocities and the penalty function of Eq. (6.19) is evaluated for whole population. The values of *pbest* are updated and the best of *pbest* is then set as *gbest* and the value of inertia weight is adjusted as per Eq. (6.23). The velocity and position are then modified according to the Eqs. (6.21) and (6.22). This completes the first iteration of PSO. The PSO algorithm runs until it reaches the maximum number of iterations. In this work, the maximum number of iterations was set to 1000. The values of C_1, C_2, w_{max}, and w_{min} were set to 2, 2, 1.2, and 0.2, respectively.

The performances of GA and PSO are compared for engineering design of fabrics. The values of optimum fabric parameters and obtained fabric properties by using GA and PSO are shown in Tables 6.16 and 6.17, respectively. The results show that the PSO is better suited for this type of application in all ranges of fabric type, achieving GSM of 72.7, 114.8, and 203.1 g/m² for light-, medium-, and heavy-weight fabrics, respectively, in comparison to 74.9, 125.3, and 213.8 g/m² as obtained with GA.

6.4 Ant colony optimization (ACO)

Inspired by the ants' capability of finding the shortest route from the nest to a food source, Dorigo et al. (2006) and Dorigo and Stützle (2019) developed ACO, which has many successful applications in discrete optimization

TABLE 6.16 Optimum values of fabric governing parameters using GA and PSO.

Fabric type	Optimization technique	T_1, tex	T_2, tex	c_1, fraction	c_2, fraction	p_1, mm	p_2, mm
Light weight	GA	13	14	0.103	0.101	0.398	0.396
	PSO	13	13	0.100	0.110	0.400	0.390
Medium weight	GA	48	43	0.040	0.062	0.784	0.741
	PSO	32	32	0.109	0.113	0.621	0.618
Heavy weight	GA	80	77	0.094	0.110	0.816	0.802
	PSO	72	72	0.117	0.113	0.789	0.792

TABLE 6.17 Obtained values of fabric physical and mechanical properties using GA and PSO.

Fabric type	Technique	K	$E_1,$ mN/ mm	$E_2,$ mN/ mm	$G,$ mN/ mm	$B_1,$ mN mm	$B_2,$ mN mm	$W,$ g/m^2
Light weight	GA	0.57	1843	1909	149.2	3.0	3.2	74.9
	PSO	0.57	1848	1718	145.3	3.0	3.0	72.7
Medium weight	GA	0.55	3096	2113	105.7	6.2	6.0	125.3
	PSO	0.57	2381	2293	200.7	6.0	6.0	114.8
Heavy weight	GA	0.65	4236	3582	309.8	12.6	12.6	213.8
	PSO	0.64	3502	3640	316.2	12.0	12.0	203.1

problems. Social insects such as ants act as a community to solve complex problems emerging in their daily lives through mutual cooperation. Ants repeatedly hop from one place to another and eventually find the shortest path to the food source. In the process of tracing a path, ants deposit an organic compound called pheromone on the ground. Ants can smell pheromones; thus, a pheromone trail is formed, which assists the ants in communicating with one another. Intensity of pheromone accumulation is higher in the shorter path and ants choose the path marked by strong pheromone concentration.

In the ACO algorithm, the task of each artificial ant is to find the shortest route between a pair of nodes on a graph that maps the problem representation. Suppose that $G=(N,A)$ is a connected graph as shown in Fig. 6.17 where A number of undirected arcs are connecting N number of nodes. A solution is a path connecting a source node S and a destination node D. The path length is constituted by the number of loops formed by each arc (i,j) of the graph. Each

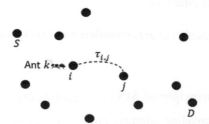

FIG. 6.17 Graph $(G=N, A)$ showing source (S) and destination (D) nodes.

arc (i,j) has an associated variable $\tau_{i,j}$ called pheromone trail intensity. The intensity of a pheromone trail is an indicator of the utility of that arc to build good solutions. At the beginning an initial value of pheromone trail intensity $\tau_{i,j}(0) = 1, \forall (i,j) \in A$ is assigned to each arc (i,j). At each node, its outgoing arc is used in a stochastic way by the ant to decide the next node to move to. The probability of the kth ant located in node i choosing node j as the next node to move to is given by

$$
P_{i,j}(k) = \begin{cases} \dfrac{\tau_{i,j}{}^{\alpha} \times \eta_{i,j}^{\beta}}{\displaystyle\sum_{k \in \text{allowed}_k} \tau_{i,j}{}^{\alpha} \times \eta_{i,j}^{\beta}} & \text{if } j \in \text{allowed}_k \\ 0 & \text{otherwise} \end{cases} \tag{6.24}
$$

where $\tau_{i,j}$ is the amount of pheromone trail present between the edge (i,j), $\eta_{i,j}$ is the visibility level between the two nodes i and j which is determined by taking the inverse of their Euclidean distance $d_{i,j}$, i.e., $\eta_{i,j} = \frac{1}{d_{i,j}}$, α is the relative importance of the pheromone trail ($\alpha \geq 0$), β is the relative importance of the visibility ($\beta \geq 0$), and allowed$_k$ is the set of available nodes from ith node to jth node from which kth ant can choose. Using the decision rule of Eq. (6.24), ants hop from the source to the destination.

In the next iteration $(t+1)$, the pheromone trail is updated by

$$
\tau_{i,j}(t+1) = (1 - \rho)\tau_{i,j}(t) + \Delta\tau_{i,j}(t, t+1) \tag{6.25}
$$

where $0 \leq \rho \leq 1$ and ρ represents the pheromone evaporation coefficient, and $\Delta\tau_{i,j}$ is expressed by

$$
\Delta\tau_{i,j} = \sum_{k=1}^{m} \Delta\tau_{i,j}(k)
$$

and

$$
\Delta\tau_{i,j}(k) = \begin{cases} 1/L(k), & \text{if } k\text{th ant uses arc } (i,j) \text{ in its tour} \\ & \text{(between iteration } t \text{ and } t+1) \\ 0, \text{otherwise} \end{cases} \tag{6.26}
$$

where m is the total number of ants considered and $L(k)$ is the length of a tour by kth ant.

6.4.1 Working principle of ACO in step-by-step

Suppose we have a traveling salesman problem (TSP) with 5 cities. Given a set of 5 cities and the distance between every pair of cities, our objective is to find

TABLE 6.18 Distance between every pair of cities.

	C_1	C_2	C_3	C_4	C_5
C_1	0	10	15	8	12
C_2	10	0	20	25	30
C_3	15	20	0	40	35
C_4	8	25	40	0	50
C_5	12	30	35	50	0

the shortest possible route that a salesman starting from city-1 visits every other city exactly once and returns to the starting city using the ACO algorithm. The distance between every pair of cities is given by the adjacency matrix as depicted in Table 6.18, where cities 1 to 5 are abbreviated as C_1 to C_5.

The first two iterations of this problem with two ants ($m = 2$) are explained in step-by-step as follows.

Step 1:

We determine the visibility level ($\eta_{i,j}$) between the two cities i and j by taking the inverse of their distance ($d_{i,j}$). Table 6.19 shows the values of visibility levels between the cities.

Then for each segment between cities we assign the initial pheromone value $\tau_{i,j}(0) = 1, \forall (i,j) \in A$ as shown in Table 6.20.

Step 2:

In this step, we determine the probability of moving from one city to another. As city-1 is the source city, it will become taboo to visit again, thus the level of visibility of city-1 is made to 0 (Table 6.21).

Then we calculate the probabilities of ant-1 to visit other cities from city-1 using Eq. (6.24) where we assume $\alpha = 1$ and $\beta = 2$. The computation details of

TABLE 6.19 Visibility levels between the cities.

	C_1	C_2	C_3	C_4	C_5
C_1	0	0.1	0.0667	0.125	0.0833
C_2	0.1	0	0.05	0.04	0.0333
C_3	0.0667	0.05	0	0.025	0.0286
C_4	0.125	0.04	0.025	0	0.02
C_5	0.0833	0.0333	0.0286	0.02	0

TABLE 6.20 Initial pheromone value $\tau_{i,j}(0)$ for each segment between cities.

	C_1	C_2	C_3	C_4	C_5
C_1	1	1	1	1	1
C_2	1	1	1	1	1
C_3	1	1	1	1	1
C_4	1	1	1	1	1
C_5	1	1	1	1	1

TABLE 6.21 Visibility levels between cities when city-1 is the beginning city.

	C_1	C_2	C_3	C_4	C_5
C_1	0	0.1	0.0667	0.125	0.0833
C_2	0	0	0.05	0.04	0.0333
C_3	0	0.05	0	0.025	0.0286
C_4	0	0.04	0.025	0	0.02
C_5	0	0.0333	0.0286	0.02	0

probabilities are depicted in Table 6.22. The last column of Table 6.22 shows the values of cumulative probabilities.

We generate a random number r to determine the choice of cities to be visited by comparing it with the value of the cumulative probability. Suppose that the generated random number is $r = 0.8147$. As $0.8124 < 0.8147 < 1$, the next place to visit is city-5. Thus city-5 will also become taboo to visit again.

TABLE 6.22 Probabilities of ant-1 going from city-1 to other cities.

From	To	$\tau_{1,j}^{\alpha} \times \eta_{1,j}^{\beta}$	$P_{1,j}(1) = \dfrac{\tau_{1,j}^{\alpha} \times \eta_{1,j}^{\beta}}{\sum \tau_{1,j}^{\alpha} \times \eta_{1,j}^{\beta}}$	Cumulative probability
C_1	C_2	$1^1 \times 0.1^2 = 0.01$	0.2702	0.2702
C_1	C_3	$1^1 \times 0.0667^2 = 0.0044$	0.1201	0.3902
C_1	C_4	$1^1 \times 0.125^2 = 0.0156$	0.4221	0.8124
C_1	C_5	$1^1 \times 0.0833^2 = 0.0069$	0.1876	1

TABLE 6.23 Visibility levels between the cities when city-5 is selected.

	C_1	C_2	C_3	C_4	C_5
C_1	0	0.1	0.0667	0.125	0
C_2	0	0	0.05	0.04	0
C_3	0	0.05	0	0.025	0
C_4	0	0.04	0.025	0	0
C_5	0	0.0333	0.0286	0.02	0

We make its visibility to 0 by setting the 5th column of the matrix equal to 0 (Table 6.23).

Next, we calculate the probabilities of ant-1 to visit other cities from city-5. The values of probabilities and cumulative probabilities are shown in Table 6.24.

Suppose we generate a random number $r=0.9058$. Because $0.8281 < 0.9058 < 1$, the next selected city is city-4. Thus, for ant-1, the route formed so far is 1-5-4 and the visibility levels between cities are shown in Table 6.25.

TABLE 6.24 Probabilities of ant-1 going from city-5 to other cities.

From	To	$\tau_{5,j}^{\alpha} \times \eta_{5,j}^{\beta}$	$P_{5,j}(1) = \dfrac{\tau_{5,j}^{\alpha} \times \eta_{5,j}^{\beta}}{\sum \tau_{5,j}^{\alpha} \times \eta_{5,j}^{\beta}}$	Cumulative probability
C_5	C_2	$1^1 \times 0.033^2 = 0.0011$	0.4774	0.4774
C_5	C_3	$1^1 \times 0.0286^2 = 0.0008$	0.3507	0.8281
C_5	C_4	$1^1 \times 0.02^2 = 0.0004$	0.1719	1

TABLE 6.25 Visibility levels between cities when the route becomes 1-5-4.

	C_1	C_2	C_3	C_4	C_5
C_1	0	0.1	0.0667	0	0
C_2	0	0	0.05	0	0
C_3	0	0.05	0	0	0
C_4	0	0.04	0.025	0	0
C_5	0	0.0333	0.0286	0	0

TABLE 6.26 Probabilities of ant-1 going from city-4 to remaining cities.

From	To	$\tau_{4,j}^{\alpha} \times \eta_{4,j}^{\beta}$	$P_{4,j}(1) = \frac{\tau_{4,j}^{\alpha} \times \eta_{4,j}^{\beta}}{\sum \tau_{4,j}^{\alpha} \times \eta_{4,j}^{\beta}}$	Cumulative probability
C_4	C_2	$1^1 \times 0.04^2 = 0.0016$	0.7191	0.7191
C_4	C_3	$1^1 \times 0.025^2 = 0.0006$	0.2809	1

In a similar way as depicted in Tables 6.22 and 6.24, the probabilities of ant-1 to visit the remaining two cities from city-4 can be obtained, which is shown in Table 6.26.

Suppose a random number $r = 0.127$ is generated. Since, $0.127 < 0.7191$, the next place to visit is city-2. As the remaining city is 3, obviously it will be the next city to travel. Thus, the route formed for ant-1 is 1-5-4-2-3, having a length of the route equal to 107.

This step is repeated for all the ants. Suppose that for ant-2 we obtain the route 1-2-3-5-4. Hence, the length of the route for ant-2 is 115.

Step 3:

In each iteration, evaporation of the pheromone will occur in all possible routes. An ant deposits additional pheromones in the routes it is passing through and the total distance is adopted to update the pheromone trail $\tau_{i,j}$ using the Eqs. (6.25) and (6.26). However, in other routes that are not passed by the ant, only evaporation will occur without the addition of a pheromone. Assuming $\rho = 0.5$, the updated pheromone trail in the untraveled routes will become

$$
\begin{aligned}
\tau_{i,j}(1) &= (1 - \rho)\tau_{i,j}(0), \text{j} \notin \text{allowed}_k \\
&= (1 - 0.5) \times 1 \\
&= 0.5
\end{aligned}
$$

Ant-1 will deposit additional pheromone $\Delta\tau_{i,j} = 1/L(1) = 1/107$ in the routes it has traversed (1-5-4-2-3). For example, while ant-1 travels from city-1 to city-5, its pheromone trail $\tau_{1,5}$ will be updated as follows

$$
\begin{aligned}
\tau_{1,5}(1) &= (1 - \rho)\tau_{1,5}(0) + \Delta\tau_{1,5}(0,1) \\
&= (1 - 0.5) \times 1 + \frac{1}{107} \\
&= 0.5 + 0.0093 \\
&= 0.5093
\end{aligned}
$$

Similarly, when ant-1 will pass from city-5 to city-4, the updated pheromone trail $\tau_{5,4}$ will also become 0.5093. In this way, ant-1 forms the updated pheromone matrix as given in Table 6.27.

TABLE 6.27 Updated pheromone matrix formed by ant-1 in the 1st iteration.

	C_1	C_2	C_3	C_4	C_5
C_1	0.5	0.5	0.5	0.5	0.5093
C_2	0.5	0.5	0.5093	0.5	0.5
C_3	0.5	0.5	0.5	0.5	0.5
C_4	0.5	0.5093	0.5	0.5	0.5
C_5	0.5	0.5	0.5	0.5093	0.5

Then ant-2 will provide additional pheromone by $1/L(2) = 1/115$ in the route it has traversed (1-2-3-5-4). For example, while ant-2 traverses from city-1 to city-2, its pheromone trail $\tau_{1,2}$ will be updated as follows

$$\tau_{1,2}(1) = 0.5 + 1/115$$
$$= 0.5 + 0.0087$$
$$= 0.5087$$

Similarly, when ant-2 will pass from city-2 to city-3, the updated pheromone trail $\tau_{2,3}$ will become

$$\tau_{2,3}(1) = 0.5093 + 1/115$$
$$= 0.5093 + 0.0087$$
$$= 0.5180$$

In a similar way, ant-2 forms the updated pheromone matrix as depicted in Table 6.28.

TABLE 6.28 Updated pheromone matrix formed by ant-2 in the 1st iteration.

	C_1	C_2	C_3	C_4	C_5
C_1	0.5	0.5087	0.5	0.5	0.5093
C_2	0.5	0.5	0.5180	0.5	0.5
C_3	0.5	0.5	0.5	0.5	0.5087
C_4	0.5	0.5093	0.5	0.5	0.5
C_5	0.5	0.5	0.5	0.5180	0.5

With this step 1st iteration completes. In the next iteration, we repeat from step 2 as follows.

Step 2:

In the 2nd iteration, we again start from ant-1 by choosing city-1 as the beginning city. Thus, the level of visibility will be as same as the values given in Table 6.21. The calculation of probabilities to visit other cities from city-1 is briefed in Table 6.29.

Say a random number $r = 0.1576$ is generated, thus the next place to visit is city-2 as $0.1576 < 0.2726$. Hence, we set the 2nd column of the matrix to 0 as shown in Table 6.30.

Next, we calculate the probabilities of ant-1 to visit other cities from city-2 in the 2nd iteration, and its values are given in Table 6.31.

This time if a random number $r = 0.9706$ is generated, the next place to visit is city-5. Therefore, 5th column of the matrix is set to 0 as shown in Table 6.32.

TABLE 6.29 Probabilities of ant-1 going from city-1 to other cities in 2nd iteration.

From	To	$\tau_{1,j}^{\alpha} \times \eta_{1,j}^{\beta}$	$P_{1,j}(1) = \frac{\tau_{1,j}^{\alpha} \times \eta_{1,j}^{\beta}}{\sum \tau_{1,j}^{\alpha} \times \eta_{1,j}^{\beta}}$	Cumulative probability
C_1	C_2	$0.5087^1 \times 0.1^2 = 0.0051$	0.2726	0.2726
C_1	C_3	$0.5^1 \times 0.0667^2 = 0.0022$	0.1191	0.3917
C_1	C_4	$0.5^1 \times 0.125^2 = 0.0078$	0.4187	0.8104
C_1	C_5	$0.5093^1 \times 0.0833^2 = 0.0035$	0.1896	1

TABLE 6.30 Visibility level between cities when city-2 is selected.

	C_1	C_2	C_3	C_4	C_5
C_1	0	0	0.0667	0.125	0.0833
C_2	0	0	0.05	0.04	0.0333
C_3	0	0	0	0.025	0.0286
C_4	0	0	0.025	0	0.02
C_5	0	0	0.0286	0.02	0

TABLE 6.31 Probabilities of ant-1 going from city-2 to other cities in 2nd iteration.

From	To	$\tau_{2,j}^{\alpha} \times \eta_{2,j}^{\beta}$	$P_{2,j}(1) = \frac{\tau_{2,j}^{\alpha} \times \eta_{2,j}^{\beta}}{\sum \tau_{2,j}^{\alpha} \times \eta_{2,j}^{\beta}}$	Cumulative probability
C_2	C_3	$0.518^1 \times 0.05^2 = 0.0013$	0.4886	0.4886
C_2	C_4	$0.5^1 \times 0.04^2 = 0.0008$	0.3018	0.7904
C_2	C_5	$0.5^1 \times 0.0333^2 = 0.0006$	0.2096	1

TABLE 6.32 Visibility level between cities when the route becomes 1-2-5.

	C_1	C_2	C_3	C_4	C_5
C_1	0	0	0.0667	0.125	0
C_2	0	0	0.05	0.04	0
C_3	0	0	0	0.025	0
C_4	0	0	0.025	0	0
C_5	0	0	0.0286	0.02	0

TABLE 6.33 Probabilities of ant-1 going from city-5 to other cities in 2nd iteration.

From	To	$\tau_{5,j}^{\alpha} \times \eta_{5,j}^{\beta}$	$P_{5,j}(1) = \frac{\tau_{5,j}^{\alpha} \times \eta_{5,j}^{\beta}}{\sum \tau_{5,j}^{\alpha} \times \eta_{5,j}^{\beta}}$	Cumulative probability
C_5	C_3	$0.5^1 \times 0.0286^2 = 0.0004$	0.6633	0.6633
C_5	C_4	$0.518^1 \times 0.02^2 = 0.0002$	0.3367	1

Similarly, the probabilities of ant-1 to visit the remaining two cities from city-5 in the 2nd iteration can be estimated, which is displayed in Table 6.33.

If the random number $r = 0.9572$, the next place to travel is city-4. Then the remaining city-3 will be the next place to visit. Thus, in the 2nd iteration, ant-1 forms the route in sequence 1-2-5-4-3. We assume that in the 2nd iteration ant-2 follows the route 1-2-3-4-5. Thus, the lengths of the route are 130 and 120 for ant 1 and ant 2, respectively.

Step 3:

In the 2nd iteration, further evaporation of pheromone trail takes place, thus the updated pheromone trail in the untraveled routes becomes

$$\tau_{i,j}(2) = (1 - \rho)\tau_{i,j}(1), j \notin \text{allowed}_k$$

$$= (1 - 0.5) \times 0.5$$

$$= 0.25$$

Ant-1 will deposit additional pheromone $\Delta\tau_{i,j} = 1/L(1) = 1/130$ in the routes it has traversed (1-2-5-4-3). For example, while ant-1 goes from city-1 to city-2, its pheromone trail $\tau_{1,2}$ is updated as follows

$$\tau_{1,2}(2) = (1 - \rho)\tau_{1,2}(1) + \Delta\tau_{1,2}(1,2)$$

$$= (1 - 0.5) \times 0.5087 + \frac{1}{130}$$

$$= 0.2543 + 0.0077$$

$$= 0.262$$

Similarly, when ant-1 passes from city-2 to city-5, the pheromone trail $\tau_{2,5}$ updates as follows

$$\tau_{2,5}(2) = (1 - \rho)\tau_{2,5}(1) + \Delta\tau_{2,5}(1,2)$$

$$= (1 - 0.5) \times 0.5 + \frac{1}{130}$$

$$= 0.25 + 0.0077$$

$$= 0.2577$$

In this way, in the 2nd iteration ant-1 forms the updated pheromone matrix as depicted in Table 6.34.

TABLE 6.34 Updated pheromone matrix formed by ant-1 in the 2nd iteration.

	C_1	C_2	C_3	C_4	C_5
C_1	0.25	0.262	0.25	0.25	0.2547
C_2	0.25	0.25	0.259	0.25	0.2577
C_3	0.25	0.25	0.25	0.25	0.2543
C_4	0.25	0.2547	0.2577	0.25	0.25
C_5	0.25	0.25	0.25	0.2667	0.25

Then ant-2 will provide additional pheromone by $1/L(2) = \frac{1}{120}$ in the route it has traversed (1-2-3-4-5). For ant-2 while traversing from city-1 to city-2, the pheromone trail $\tau_{1,2}$ will be updated as follows

$$\tau_{1,2}(2) = 0.262 + 1/120$$
$$= 0.262 + 0.0083$$
$$= 0.2703$$

In a similar way, in the 2nd iteration ant-2 forms the updated pheromone matrix as shown in Table 6.35.

With this step, 2nd iteration is completed. Then we proceed to the next iteration with step 2 again. Steps 2–3 will be repeated until the maximum iteration is reached. MATLAB coding for solving this traveling salesman problem using ACO is presented in Section 6.5 where the number of ants and maximum iteration are considered to be 100 and 1000, respectively. Table 6.36 depicts the

TABLE 6.35 Updated pheromone matrix formed by ant-2 in the 2nd iteration.

	C_1	C_2	C_3	C_4	C_5
C_1	0.25	0.2703	0.25	0.25	0.2547
C_2	0.25	0.25	0.2673	0.25	0.2577
C_3	0.25	0.25	0.25	0.2583	0.2543
C_4	0.25	0.2547	0.2577	0.25	0.2583
C_5	0.25	0.25	0.25	0.2667	0.25

TABLE 6.36 Updated pheromone matrix after the end of 1000 iterations by 100 ants.

	C_1	C_2	C_3	C_4	C_5
C_1	0	0	0	2.2727	0
C_2	0	0	2.2727	0	0
C_3	0	0	0	0	2.2727
C_4	0	2.2727	0	0	0
C_5	0	0	0	0	0

updated pheromone matrix after the end of maximum iterations by all ants. It is evident from Table 6.36 that the best route is 1-4-2-3-5 having a route length of 88. As the salesman returns to the starting city-1, the optimized route becomes 1-4-2-3-5-1.

6.4.2 Application of ACO

The traveling salesman problem (TSP) has many real-world applications, including planning delivery routes. TSP involves finding the shortest possible route that a salesperson visits a set of cities and returns to the starting point, with the constraint that each city must be visited exactly once. Let us imagine that a garment-making industry needs to supply its products in 10 different cities of India. The distances in Km between each city are given in Table 6.37 The salesperson starting from Ahmedabad wants to visit each city exactly once and return to the starting point while traveling the shortest distance possible. A possible solution may be assumed to be the following route:

$$1 - 2 - 3 - 4 - 5 - 6 - 7 - 8 - 9 - 10 - 1.$$

The total distance of this route is estimated as

$524 + 1270 + 332 + 2154 + 1562 + 1470 + 1010 + 1330 + 1230 + 390$
$= 11272 \text{ km}$

However, this may not be the optimal solution. Theoretically, there are a total of $10! = 3628800$ possible routes. We need to find the shortest possible route using the ACO algorithm.

To solve the problem, we assumed the ACO parameters are as follows:

Number of ants $= 100$, evaporation rate $= 0.5, \alpha = 1$,
$\beta = 2$, and number of maximum iterations $= 1000$.

Table 6.38 shows the values of visibility levels among the cities that are obtained by taking the inverse of the distance between any two cities. Using, the ACO algorithm, the best route is found to be

$$1 - 10 - 2 - 6 - 4 - 3 - 9 - 7 - 8 - 5 - 1.$$

Fig. 6.18 illustrates a schematic diagram of the shortest route map. The total length of this shortest route is estimated to be

$390 + 585 + 716 + 570 + 332 + 795 + 880 + 1010 + 534 + 935 = 6747 \text{ km}$

Table 6.39 shows the final pheromone matrix at the end of 1000 iterations.

TABLE 6.37 The distance between the cities.

	Ahmadabad (1)	Mumbai (2)	Chennai (3)	Bengaluru (4)	Delhi (5)	Hyderabad (6)	Kolkata (7)	Lucknow (8)	Visakhapatnam (9)	Indore (10)
Ahmadabad (1)	0	524	1760	1495	935	1186	2020	1212	1660	390
Mumbai (2)	524	0	1270	985	1430	716	1900	1380	1345	585
Chennai (3)	1760	1270	0	332	2188	627	1665	1975	795	1485
Bengaluru (4)	1495	985	332	0	2154	570	2048	2008	975	1330
Delhi (5)	935	1430	2188	2154	0	1562	1550	534	1766	777
Hyderabad (6)	1186	716	627	570	1562	0	1470	1280	590	792
Kolkata (7)	2020	1900	1665	2048	1550	1470	0	1010	880	1562
Lucknow (8)	1212	1380	1975	2008	534	1280	1010	0	1330	790
Visakhapatnam (9)	1660	1345	795	975	1766	590	880	1330	0	1230
Indore (10)	390	585	1485	1330	777	792	1562	790	1230	0

TABLE 6.38 Visibility levels among the cities.

	1	2	3	4	5	6	7	8	9	10
1	0	0.0019	0.0006	0.0007	0.0011	0.0008	0.0005	0.0008	0.0006	0.0026
2	0.0019	0	0.0008	0.0010	0.0007	0.0014	0.0005	0.0007	0.0007	0.0017
3	0.0006	0.0008	0	0.0030	0.0005	0.0016	0.0006	0.0005	0.0013	0.0007
4	0.0007	0.0010	0.0030	0	0.0005	0.0018	0.0005	0.0005	0.0010	0.0008
5	0.0011	0.0007	0.0005	0.0005	0	0.0006	0.0006	0.0019	0.0006	0.0013
6	0.0008	0.0014	0.0016	0.0018	0.0006	0	0.0007	0.0008	0.0017	0.0013
7	0.0005	0.0005	0.0006	0.0005	0.0006	0.0007	0	0.0010	0.0011	0.0006
8	0.0008	0.0007	0.0005	0.0005	0.0019	0.0008	0.0010	0	0.0008	0.0013
9	0.0006	0.0007	0.0013	0.0010	0.0006	0.0017	0.0011	0.0008	0	0.0008
10	0.0026	0.0017	0.0007	0.0008	0.0013	0.0013	0.0006	0.0013	0.0008	0

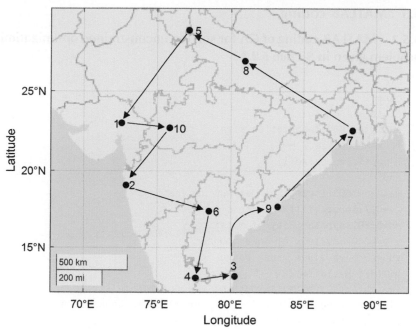

FIG. 6.18 Shortest route map.

TABLE 6.39 Final pheromone matrix.

	1	2	3	4	5	6	7	8	9	10
1	0	0	0	0	0	0	0	0	0	0.0344
2	0	0	0	0	0	0.0344	0	0	0	0
3	0	0	0	0	0	0	0	0	0.0344	0
4	0	0	0.0344	0	0	0	0	0	0	0
5	0	0	0	0	0	0	0	0	0	0
6	0	0	0	0.0344	0	0	0	0	0	0
7	0	0	0	0	0	0	0	0.0344	0	0
8	0	0	0	0	0.0344	0	0	0	0	0
9	0	0	0	0	0	0	0.0344	0	0	0
10	0	0.0344	0	0	0	0	0	0	0	0

6.5 MATLAB coding

6.5.1 MATLAB coding of GA for solving unconstrained optimization problem given in Section 6.2.3.1

```
clc
clear
close all
popsize=1000;%population size
if rem(popsize,2)==1
  popsize=popsize+1;
end
maxgen=1000;%number of maximum generations
pc=0.7;%probability of cross-over
pm=0.001;%probability of mutation
vlb=[0 0];%values of lower bound
vub=[5 5];%values upper bound
bits=[8 8];%bitsize
len_chrom=sum(bits);%Length of the chromosome/string
old_gen=round(rand(popsize,len_chrom));%Old generation
[old_gen_row,old_gen_col]=size(old_gen);
accuracy=(vub-vlb)./((2.^bits)-1);%accuracy
temp=zeros(old_gen_row,length(bits));
for i=1:old_gen_row
    bit_count=0;
    for j=1:length(bits)
        z=bit_count;
        bit_count=bit_count+bits(j);
        pow=bits(j)-1;
        z=z+1;
            while z<=bit_count
                temp(i,j)=temp(i,j)+(2^pow)*old_gen(i,z);
                pow=pow-1;
                z=z+1;
            end
        xt(i,j)=vlb(1,j)+temp(i,j)*accuracy(1,j);
    end
end
xt;%Equvalent decimal values of chromosome for each variable
for i=1:popsize
    x1=xt(i,1);
    x2=xt(i,2);
    M=(x1^2+x2-11)^2+(x1+x2^2-7)^2;%objective function
    fitness(i,1)=1/(1+M);%fitness function
end
```

```
%Roulette wheel selection method
for gct=1:maxgen
    gct%generation count
    sum_fit=sum(fitness);
    ave_fit=sum_fit/popsize;
    A=fitness./(ave_fit);
    B=A/popsize;%probability
    for i=1:popsize
        if i==1
        C(i,1)=B(i);
        else
        C(i,1)=C(i-1,1)+B(i);%cumulative probability
        end
    end
    D=rand(popsize,1);%random number between 0 and 1
    for i=1:popsize
        sel_index(i)=min(find(D(i,1)<C(:,1)));
    end
    E=sel_index';%selected index
    for i=1:popsize
        str_count(i)=0;
        for j=1:length(E)
            if sel_index(j)==i
            str_count(i)=str_count(i)+1;
            end
        end
    end
    F=str_count';%actual count in the mating pool
    selected_individual=find(F);
    copy=nonzeros(F);
    roultte_wheel_selection=[selected_individual copy];
    sel_index;
    old_gen;
    new_gen=old_gen(sel_index,:);
%Cross over (uniform cross-over)
    lchrom=size(new_gen,2);
    chrom_number=size(new_gen,1);
    ma=ceil(rand(size(new_gen,1)/2,lchrom)-pc);
    for i=1:size(new_gen,1)/2
        for j=1:lchrom
            if ma(i,j)==0
            new_gen_x(2*i-1,j)=new_gen(2*i-1,j);
            new_gen_x(2*i,j)=new_gen(2*i,j);
```

```
                else
                new_gen_x(2*i-1,j)=new_gen(2*i,j);
                new_gen_x(2*i,j)=new_gen(2*i-1,j);
                end
            end
        end
%Mutation
    size(new_gen_x);
    r=rand(size(new_gen_x));
    mutated=find(r<pm);
    new_gen_m=new_gen_x;
    new_gen_m(mutated)=1-new_gen_x(mutated);
%Fitness evaluation
    [new_gen_m_row,new_gen_m_col]=size(new_gen_m);
    acy=(vub-vlb)./((2.^bits)-1);
    te=zeros(new_gen_m_row,length(bits));
    for i=1:new_gen_m_row
        bit_ct=0;
        for j=1:length(bits)
            q=bit_ct;
            bit_ct=bit_ct+bits(j);
            pw=bits(j)-1;
            q=q+1;
            while q<=bit_ct
                    te(i,j)=te(i,j)+(2^pw)*new_gen_m(i,q);
                    pw=pw-1;
                    q=q+1;
            end
            xt_new(i,j)=vlb(1,j)+te(i,j)*acy(1,j);
        end
    end
    for i=1:popsize
        x1=xt_new(i,1);
        x2=xt_new(i,2);
        M=(x1^2+x2-11)^2+(x1+x2^2-7)^2;%objective function
        new_fitness(i,1)=1/(1+M);%fitness function
    end
    ave=sum(new_fitness)/popsize;
    diff=abs(new_fitness-ave);
    jj=min(find(new_fitness==max(new_fitness)));
    ratio=ave/new_fitness(jj)%ratio
    for j=1:length(bits)
        best_fit_values(j)=xt_new(jj,j);
    end
```

```
    if ratio>=0.99
    break
    end
    fitness=new_fitness;
    old_gen=new_gen_m;
end
x1=best_fit_values(1,1)
x2=best_fit_values(1,2)
M=(x1^2+x2-11)^2+(x1+x2^2-7)^2%optimum value of function
```

6.5.2   MATLAB coding of PSO for solving optimization problem given in Section 6.3.2

```
clc
clear
close all
format short g
m=2; % number of variables
n=1000; % swarm size
w=0.5;%Inertia weight
c1=2;%Learning factor 1
c2=2;%Learning factor 2
ub=[5 5];%Upper bound of variables
lb=[0 0];%Lower bound of variables
itr=100;%Number of iterations
%Initialize the swarm position and swarm velocity
for i=1:n
    for j=1:m
        x0(i,j)=(lb(j)+rand()*(ub(j)-lb(j)));
    end
end
v=0.1*x0; % initial velocity
fx0=(x0(:,1).^2+x0(:,2)-11).^2+(x0(:,1)+x0(:,2).^2-7).^2;
%Initial fitness evaluation
pbest=x0; % initial pbest
[fmin0,index0]=min(fx0);
gbest=x0(index0,:); % initial gbest
x=x0; % initial population
% PSO starts.....
for j=1:itr
    % pso velocity update
    for i=1:n
        for j=1:m
            v(i,j)=w*v(i,j)+c1*rand()*(pbest(i,j)-x(i,j))...
```

```
                    +c2*rand()*(gbest(1,j)-x(i,j));
            end
        end
        % pso position update
        for i=1:n
            for j=1:m
                x(i,j)=x(i,j)+v(i,j);
            end
        end
        % handling boundary violations
        for i=1:n
            for j=1:m
                if x(i,j)<lb(j)
                    x(i,j)=lb(j);
                elseif x(i,j)>ub(j)
                    x(i,j)=ub(j);
                end
            end
        end
        fx=(x(:,1).^2+x(:,2)-11).^2+(x(:,1)+x(:,2).^2-7).^2;%
    fitness function
        % updating pbest and fitness
        for i=1:n
            if fx(i,1)<fx0(i,1)
                pbest(i,:)=x(i,:);
                fx0(i,1)=fx(i,1);
            end
        end
        [fmin,index]=min(fx0); % finding out the best particle
        %updating gbest and best fitness
        if fmin<fmin0
            gbest=pbest(index,:);
            fmin0=fmin;
        end
    end
    x=gbest
    fn_val=(gbest(:,1).^2+gbest(:,2)-11).^2+(gbest(:,1)+gbest
    (:,2).^2-7).^2
```

6.5.3 MATLAB coding of ACO for solving optimization problem given in Section 6.4.2

```
clc
clear
```

```
close all
format short
x=[0 10 15 8 12
   10 0 20 25 30
   15 20 0 40 35
   8  25 40 0 50
   12 30 35 50 0];
n=length(x);%number of cities
nm=100;%number of ant
a=1;%alpha
b=2;%beta
evp=0.5;%evaporation rate
max_iter=1000;%maximum iteration
init_tao=ones(n);%initialization
for i=1:n
    for j=1:n
        if x(i,j)==0
            y(i,j)=0;
        else
            y(i,j)=1/x(i,j);
        end
    end
end
for itr=1:max_iter
    for i=1:nm
        route(i,1)=1;
    end
    for i=1:nm
        cy=y;
        for j=1:n
            c=route(i,j);
            cy(:,c)=0;
            p=(init_tao(c,:).^a).*(cy(c,:).^b);
            sm=(sum(p));
            if sm==0
                pr=0*p;
            else
                pr=(1/sm).*p;
            end
            p;
            pr;
            cpr=cumsum(pr);
            r=rand;
```

```
            for q=1:n
                if r<=cpr(q)
                    route(i,j+1)=q;
                    selected_route=q;
                break
                end
            end
        end
    end
    route;
    c_route=horzcat(route,route(:,1));
    for i=1:nm
        s=0;
        for j=1:n-1
            s=s+x(c_route(i,j),c_route(i,j+1));
        end
        L(i)=s;
    end
    L;
    [Lmin,I]=min(L);
    best_route=c_route(I,:);
    init_tao=(1-evp)*init_tao;
    for i=1:nm
        for j=1:n-1
            d_tao=1/L(i);
init_tao(c_route(i,j),c_route(i,j+1))=init_tao(c_route(i,j),
c_route(i,j+1))+d_tao;
        end
    end
end
init_tao
best_route
Lmin
```

6.6 Summary

This chapter delves into the popularly known nature-inspired optimization algorithms, such as GA, PSO, and ACO. For each method, the underlying mechanisms and key components are explained, accompanied by illustrative examples in the textile domain to provide a clear understanding of their functioning. This chapter offers a comprehensive resource for researchers, practitioners, and students interested in understanding and applying nature-inspired methods in solving complex optimization problems across different domains.

References

Davis, L. (1991). *Handbook of genetic algorithms*. New York: Van Nostrand, Reinhold.

Deb, K. (2005). *Optimization for engineering design: Algorithms and examples*. New Delhi, India: Prentice Hall of India Private Limited.

Dorigo, M., Birattari, M., & Stutzle, T. (2006). Ant colony optimization. *IEEE Computational Intelligence Magazine, 1*(4), 28–39.

Dorigo, M., & Stützle, T. (2019). *Ant Colony optimization: Overview and recent advances*. Springer International Publishing.

Ghosh, A., Mal, P., & Majumdar, A. (2019). *Advanced optimization and decision-making techniques in textile engineering*. Boca Raton: CRC Press/Taylor & Francis Group.

Goldberg, D. E. (1989). *Genetic algorithms in search, optimization & machine learning* (1st ed.). New York: Addison-Wesley.

Holland, J. H. (1975). *Adaptation in natural and artificial system*. Ann Arbor: The University of Michigan Press.

Kennedy, J., & Eberhart, R. C. (1995). Particle swarm optimization. In *Proceedings of IEEE international conference on neural networks*, Perth, Australia (pp. 1942–1948).

Leaf, G. A. V. (2003). The mechanics of plain woven fabrics. In *Proceedings of international textile design and engineering conference INTEDEC*, Edinburgh, UK.

Leaf, G. A. V., Chen, Y., & Chen, X. (1993). The initial bending behaviour of plain woven fabrics. *Journal of the Textile Insitute, 84*(3), 419–428.

Leaf, G. A. V., & Kandil, K. H. (1980). The initial load-extension behaviour of plain-woven fabric. *Journal of the Textile Insitute, 71*(1), 1–7.

Leaf, G. A. V., & Sheta, A. M. F. (1984). The initial shear modulus of plain-woven fabrics. *Journal of the Textile Insitute, 75*(3), 157–163.

Morton, W. E., & Hearle, J. W. S. (1993). *Physical properties of textile fibres* (3rd ed.). Manchester: The Textile Institute.

Peirce, F. T. (1937). The geometry of cloth structure. *Journal of the Textile Institute Transaction, 28*(3), 45–112.

Postle, R., Caranby, G. A., & de Jong, S. (1988). *The mechanics of wool structures*. Chichester: Ellis Horwood.

Shi, Y., & Eberhart, R.C. (1998a). A modified particle swarm optimizer. Proceedings of IEEE international conference on computational intelligence, Anchorage, AK (pp. 69–73).

Shi, Y., & Eberhart, R. C. (1998b). Parameter selection in particle swarm adaptation. In V. W. Porto, N. Saravanan, D. Waagen, & A. E. Eiben (Eds.), *Evolutionary programming VII* (pp. 591–600). Berlin: Springer.

Umapathy, P., Venkataseshaiah, C., & Senthil, M. (2010). Particle swarm optimization with various inertia weight variants for optimal power flow solution. *Discrete Dynamics in Nature and Society, 10*, 1–15.

Chapter 7

Hybrid artificial intelligence systems

7.1 Introduction

In the domain of artificial intelligence (AI), the emergence of hybrid artificial intelligence systems opens new avenues by combining the capabilities of at least two different AI techniques to create more robust, adaptable, comprehensive, and high-performing AI systems. In the previous chapters we have discussed that fuzzy logic is renowned for its ability to capture imprecision and uncertainty in human reasoning, artificial neural networks (ANNs) excel at learning patterns from data through their interconnected layers of artificial neurons and genetic algorithms are powerful optimization techniques inspired by the process of natural evolution.

Adaptive neuro-fuzzy inference systems (ANFIS) represent a powerful and versatile approach that fuses the strengths of fuzzy logic and ANNs (Jang, 1993; Jang et al., 1997). By combining the fuzzy logic's linguistic rule-based approach with the data-driven learning capabilities of ANNs, ANFIS bridges the gap between symbolic reasoning and statistical learning.

A genetic fuzzy expert system (GFES) is a hybrid intelligent approach that combines the power of genetic algorithms and fuzzy expert systems to handle complex, uncertain, and dynamic problems more effectively by learning from the data. In a GFES, genetic algorithm optimizes the fuzzy rules and membership functions within a fuzzy expert system.

This chapter provides a comprehensive introduction to the concept of two hybrid artificial intelligence systems, namely ANFIS and GFES by elucidating their fundamental principles, components, and applications.

7.2 ANFIS

When a fuzzy inference system (FIS) based on Sugeno's approach is represented using the structure of an ANN and trained by the combination of backpropagation algorithm and least squares method, the developed system is known as an ANFIS. A brief description of ANFIS is given as follows.

Artificial Intelligence in Textile Engineering. https://doi.org/10.1016/B978-0-443-15395-2.00004-1

For the purpose of simplicity, let us assume that an ANFIS model comprises of two inputs (x_1 and x_2) and a single output. Let us also suppose that, each input is represented by two linguistic terms, such as "low" (L) and "high" (H). The membership function distributions of the input variables are assumed to be bell-shaped having three parameters a, b, and c. As there are two inputs and each of them has two linguistic terms, there is a maximum of $2^2 = 4$ number of rules. According to Sugeno's method of fuzzy logic system, the output of each rule can be expressed as:

$$y_i = \beta_{i0} + \beta_{i1}x_1 + \beta_{i2}x_2 \tag{7.1}$$

where $i = 1, 2, 3, 4$ and $\beta_{i0}, \beta_{i1}, \beta_{i2}$ are the coefficients. To solve this problem, an ANFIS architecture shown in Fig. 7.1 may be considered, which consists of six layers. In the event that the following four rules are triggered:

If x_1 is low and x_2 is low, then $y_1 = \beta_{10} + \beta_{11}x_1 + \beta_{12}x_2$,

If x_1 is low and x_2 is high, then $y_2 = \beta_{20} + \beta_{21}x_1 + \beta_{22}x_2$,

If x_1 is low and x_2 is high, then $y_2 = \beta_{30} + \beta_{31}x_1 + \beta_{32}x_2$,

If x_1 is high and x_2 is high, then $y_4 = \beta_{40} + \beta_{41}x_1 + \beta_{42}x_2$.

To represent the inputs and outputs of a layer R, we use the notations $x_j^{(R)}$ and $y_j^{(R)}$ which indicate the input and output, respectively, of jth node lying on Rth layer ($R = 1, 2, ..., 6$). An ANFIS architecture consists of 6 layers. The inputs and outputs of each layer are discussed as follows.

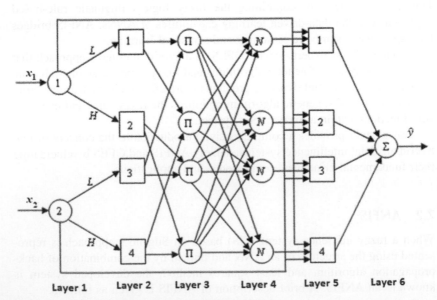

FIG. 7.1 Schematic representation of an ANFIS architecture.

Layer 1: It is the input layer of the network consisting of two nodes. The function of this layer is to pass the inputs x_1 and x_2 to the next layer. The outputs of the nodes are the same as the corresponding inputs, i.e.,

$$y_1^{(1)} = x_1^{(1)} = x_1$$
$$y_2^{(1)} = x_2^{(1)} = x_2$$

Layer 2: It determines the membership values for a set of inputs corresponding to their appropriate linguistic terms. The inputs of this layer are

$$x_1^{(2)} = x_1$$
$$x_2^{(2)} = x_1$$
$$x_3^{(2)} = x_2$$
$$x_4^{(2)} = x_2$$

The corresponding outputs are

$$y_1^{(2)} = \frac{1}{1 + \left| \frac{x_1^{(2)} - c_1}{a_1} \right|^{2b_1}} = \mu_L(x_1)$$

$$y_2^{(2)} = \frac{1}{1 + \left| \frac{x_2^{(2)} - c_2}{a_2} \right|^{2b_2}} = \mu_H(x_1)$$

$$y_3^{(2)} = \frac{1}{1 + \left| \frac{x_3^{(2)} - c_3}{a_3} \right|^{2b_3}} = \mu_L(x_2)$$

$$y_4^{(2)} = \frac{1}{1 + \left| \frac{x_4^{(2)} - c_4}{a_4} \right|^{2b_4}} = \mu_H(x_2)$$

Layer 3: It contains 4 nodes, which are assigned by the symbol Π. Each node indicates a possible combination of the input variables. All nodes are activated corresponding to the above four fired rules. The output of every node of this layer is the product of all incoming signals, which is also known as the firing strength (μ_j) of the corresponding node. The inputs of this layer are

$$x_1^{(3)} = \left[y_1^{(2)}, y_3^{(2)} \right]$$
$$x_2^{(3)} = \left[y_1^{(2)}, y_4^{(2)} \right]$$
$$x_3^{(3)} = \left[y_2^{(2)}, y_3^{(2)} \right]$$
$$x_4^{(3)} = \left[y_2^{(2)}, y_4^{(2)} \right]$$

Hence, the corresponding outputs are

$$y_1^{(3)} = y_1^{(2)} y_3^{(2)} = \mu_L(x_1)\mu_L(x_2) = \mu_1$$

$$y_2^{(3)} = y_1^{(2)} y_4^{(2)} = \mu_L(x_1)\mu_H(x_2) = \mu_2$$

$$y_3^{(3)} = y_2^{(2)} y_3^{(2)} = \mu_H(x_1)\mu_L(x_2) = \mu_3$$

$$y_4^{(3)} = y_2^{(2)} y_4^{(2)} = \mu_H(x_1)\mu_H(x_2) = \mu_4$$

Layer 4: It also contains 4 nodes as that of the previous layer, which are denoted by the symbol N. The output of each node of this layer, which is called the normalized firing strength $(\bar{\mu}_j)$, is calculated as the ratio of firing strength of that node to the sum of strengths of all fired rules. The input of each node can be expressed as:

$$x_1^{(4)} = [\mu_1, \mu_2, \mu_3, \mu_4]$$

$$x_2^{(4)} = [\mu_1, \mu_2, \mu_3, \mu_4]$$

$$x_3^{(4)} = [\mu_1, \mu_2, \mu_3, \mu_4]$$

$$x_4^{(4)} = [\mu_1, \mu_2, \mu_3, \mu_4]$$

The corresponding outputs are

$$y_1^{(4)} = \frac{\mu_1}{\mu_1 + \mu_2 + \mu_3 + \mu_4} = \bar{\mu}_1$$

$$y_2^{(4)} = \frac{\mu_2}{\mu_1 + \mu_2 + \mu_3 + \mu_4} = \bar{\mu}_2$$

$$y_3^{(4)} = \frac{\mu_3}{\mu_1 + \mu_2 + \mu_3 + \mu_4} = \bar{\mu}_3$$

$$y_4^{(4)} = \frac{\mu_4}{\mu_1 + \mu_2 + \mu_3 + \mu_4} = \bar{\mu}_4$$

Layer 5: It contains 4 nodes and all of them are activated for a set of inputs. The output of each node of this layer is calculated as the product of its normalized firing strength $(\bar{\mu}_j)$ and output of the corresponding fired rule.

The input of each node can be expressed as:

$$x_1^{(5)} = \bar{\mu}_1$$

$$x_2^{(5)} = \bar{\mu}_2$$

$$x_3^{(5)} = \bar{\mu}_3$$

$$x_4^{(5)} = \bar{\mu}_4$$

The corresponding outputs are

$$y_1^{(5)} = \bar{\mu}_1 y_1 = \bar{\mu}_1 [\beta_{10} + \beta_{11} x_1 + \beta_{12} x_2]$$

$$y_2^{(5)} = \bar{\mu}_2 y_2 = \bar{\mu}_2 [\beta_{20} + \beta_{21} x_1 + \beta_{22} x_2]$$

$$y_3^{(5)} = \bar{\mu}_3 y_3 = \bar{\mu}_3 [\beta_{30} + \beta_{31} x_1 + \beta_{32} x_2]$$

$$y_4^{(5)} = \bar{\mu}_4 y_4 = \bar{\mu}_4 [\beta_{40} + \beta_{41} x_1 + \beta_{42} x_2]$$

Layer 6: This layer consists of only one node, which is represented by the symbol Σ. It computes the overall output as the summation of all incoming signals.

$$x_1^{(6)} = [\bar{\mu}_1 y_1, \bar{\mu}_2 y_2, \bar{\mu}_3 y_3, \bar{\mu}_4 y_4]$$

Therefore, the overall predicted output ($\hat{y}$) is determined as follows:

$$\hat{y} = y_1^{(6)} = \bar{\mu}_1 y_1 + \bar{\mu}_2 y_2 + \bar{\mu}_3 y_3 + \bar{\mu}_4 y_4$$
$$= \bar{\mu}_1 [\beta_{10} + \beta_{11} x_1 + \beta_{12} x_2] + \bar{\mu}_2 [\beta_{20} + \beta_{21} x_1 + \beta_{22} x_2] +$$
$$\bar{\mu}_3 [\beta_{30} + \beta_{31} x_1 + \beta_{32} x_2] + \bar{\mu}_4 [\beta_{40} + \beta_{41} x_1 + \beta_{42} x_2]$$

The desired output (y) can be expressed as

$$y = \hat{y} + e$$
$$= \bar{\mu}_1 [\beta_{10} + \beta_{11} x_1 + \beta_{12} x_2] + \bar{\mu}_2 [\beta_{20} + \beta_{21} x_1 + \beta_{22} x_2] +$$
$$\bar{\mu}_3 [\beta_{30} + \beta_{31} x_1 + \beta_{32} x_2] + \bar{\mu}_4 [\beta_{40} + \beta_{41} x_1 + \beta_{42} x_2] + e$$

where e is the error term. For a training set of K input-output patterns, K number of linear equations are formed as follows:

$$y(1) = \bar{\mu}_1(1)[\beta_{10} + \beta_{11} x_1(1) + \beta_{12} x_2(1)] + \bar{\mu}_2(1)[\beta_{20} + \beta_{21} x_1(1) + \beta_{22} x_2(1)] +$$
$$\bar{\mu}_3(1)[\beta_{30} + \beta_{31} x_1(1) + \beta_{32} x_2(1)] + \bar{\mu}_4(1)[\beta_{40} + \beta_{41} x_1(1) + \beta_{42} x_2(1)] + e(1)$$

$$y(2) = \bar{\mu}_1(2)[\beta_{10} + \beta_{11} x_1(2) + \beta_{12} x_2(2)] + \bar{\mu}_2(2)[\beta_{20} + \beta_{21} x_1(2) + \beta_{22} x_2(2)] +$$
$$\bar{\mu}_3(2)[\beta_{30} + \beta_{31} x_1(2) + \beta_{32} x_2(2)] + \bar{\mu}_4(2)[\beta_{40} + \beta_{41} x_1(2) + \beta_{42} x_2(2)] + e(2)$$

$$\vdots$$

$$y(k) = \bar{\mu}_1(\beta)[\beta_{10} + \beta_{11} x_1(k) + \beta_{12} x_2(k)] + \bar{\mu}_2(k)[\beta_{20} + \beta_{21} x_1(k) + \beta_{22} x_2(k)] +$$
$$\bar{\mu}_3(k)[\beta_{30} + \beta_{31} x_1(k) + \beta_{32} x_2(k)] + \bar{\mu}_4(k)[\beta_{40} + \beta_{41} x_1(k) + \beta_{42} x_2(k)] + e(k)$$

$$\vdots$$

$$y(K) = \bar{\mu}_1(K)[\beta_{10} + \beta_{11} x_1(K) + \beta_{12} x_2(K)] + \bar{\mu}_2(K)[\beta_{20} + \beta_{21} x_1(K) + \beta_{22} x_2(K)] +$$
$$\bar{\mu}_3(K)[\beta_{30} + \beta_{31} x_1(K) + \beta_{32} x_2(K)] + \bar{\mu}_4(K)[\beta_{40} + \beta_{41} x_1(K) + \beta_{42} x_2(K)] + e(K)$$

Using the matrix notation, we denote

$$Y = \begin{bmatrix} y(1) \\ y(2) \\ \vdots \\ y(k) \\ \vdots \\ y(K) \end{bmatrix}$$

$$X = \begin{bmatrix} \bar{\mu}_1(1) & \bar{\mu}_1(1)x_1(1) & \bar{\mu}_1(1)x_2(1) & \bar{\mu}_2(1) & \bar{\mu}_2(1)x_1(1) & \bar{\mu}_2(1)x_2(1) & \bar{\mu}_3(1) & \bar{\mu}_3(1)x_1(1) & \bar{\mu}_3(1)x_2(1) & \bar{\mu}_4(1) & \bar{\mu}_4(1)x_1(1) & \bar{\mu}_4(1)x_2(1) \\ \bar{\mu}_1(2) & \bar{\mu}_1(2)x_1(2) & \bar{\mu}_1(2)x_2(2) & \bar{\mu}_2(2) & \bar{\mu}_2(2)x_1(2) & \bar{\mu}_2(2)x_2(2) & \bar{\mu}_3(2) & \bar{\mu}_3(2)x_1(2) & \bar{\mu}_3(2)x_2(2) & \bar{\mu}_4(2) & \bar{\mu}_4(2)x_1(2) & \bar{\mu}_4(2)x_2(2) \\ \cdots & \cdots & \cdots & \cdots & \cdots & \cdots & \cdots & \cdots & \cdots & \cdots & \cdots & \cdots \\ \bar{\mu}_1(k) & \bar{\mu}_1(k)x_1(k) & \bar{\mu}_1(k)x_2(k) & \bar{\mu}_2(k) & \bar{\mu}_2(k)x_1(k) & \bar{\mu}_2(k)x_2(k) & \bar{\mu}_3(k) & \bar{\mu}_3(k)x_1(k) & \bar{\mu}_3(k)x_2(k) & \bar{\mu}_4(k) & \bar{\mu}_4(k)x_1(k) & \bar{\mu}_4(k)x_2(k) \\ \cdots & \cdots & \cdots & \cdots & \cdots & \cdots & \cdots & \cdots & \cdots & \cdots & \cdots & \cdots \\ \bar{\mu}_1(K) & \bar{\mu}_1(K)x_1(K) & \bar{\mu}_1(K)x_2(K) & \bar{\mu}_2(K) & \bar{\mu}_2(K)x_1(K) & \bar{\mu}_2(K)x_2(K) & \bar{\mu}_3(K) & \bar{\mu}_3(K)x_1(K) & \bar{\mu}_3(K)x_2(K) & \bar{\mu}_4(K) & \bar{\mu}_4(K)x_1(K) & \bar{\mu}_4(K)x_2(K) \end{bmatrix}$$

$$B = \begin{bmatrix} \beta_{10} \\ \beta_{11} \\ \beta_{12} \\ \beta_{20} \\ \beta_{21} \\ \beta_{22} \\ \beta_{30} \\ \beta_{31} \\ \beta_{32} \\ \beta_{40} \\ \beta_{41} \\ \beta_{42} \end{bmatrix}$$

$$e = \begin{bmatrix} e(1) \\ e(2) \\ \vdots \\ e(k) \\ \vdots \\ e(K) \end{bmatrix}$$

Hence, the set of above linear equations can be written as

$$Y = XB + e$$

Or,

$$e = Y - XB$$

Now the error sum of square can be expressed as

$$\frac{1}{2} \sum_{i=1}^{K} e^2(i) = \frac{1}{2} [e(1)\ e(2)\ \cdots\ e(k)\ \ldots\ e(K)] \begin{bmatrix} e(1) \\ e(2) \\ \vdots \\ e(k) \\ \vdots \\ e(K) \end{bmatrix}$$

$$= \frac{1}{2} e^T e$$

where e^T is the transpose of e.

Thus, the error sum of square becomes

$$E = \frac{1}{2} e^T e$$

$$= \frac{1}{2} (Y - XB)^T (Y - XB)$$

$$= \frac{1}{2} \left(Y^T - B^T X^T \right)(Y - XB)$$

$$= \frac{1}{2} \left(Y^T Y - B^T X^T Y - Y^T XB + B^T X^T XB \right)$$

The partial derivative of E with respect to B gives

$$\frac{\partial E}{\partial B} = \frac{1}{2} \left(-X^T Y - Y^T X + 2X^T XB \right)$$

As $X^T Y = Y^T X$, we can write

$$\frac{\partial E}{\partial B} = -X^T Y + X^T XB$$

Setting $\frac{\partial E}{\partial B} = 0$, we get

$$X^T XB = X^T Y$$

Thus,

$$B = (X^T X)^{-1} X^T Y$$

where $(X^T X)^{-1}$ is the inverse of $X^T X$ and $(X^T X)^{-1} X^T$ is the pseudo-inverse of X if $X^T X$ is nonsingular.

For the ANFIS model, the total number of linear parameters (N_1) can be calculated as follows:

$$N_1 = (n + 1) \times p^n \tag{7.2}$$

where n is the number of inputs, p is the number of membership functions for each input. The total number of nonlinear parameters (N_2) can be estimated as follows:

$$N_2 = n \times p \times F \tag{7.3}$$

where F is the number of parameters defining the membership function.

The performance of ANFIS depends on the membership function distributions of the input variables and the parameters a_i, b_i, and c_i, which are upgraded by the learning process as follows:

Using the chain rule, we can write

$$\Delta a_i = -\alpha \frac{\partial E}{\partial a_i}$$

$$= -\alpha \frac{\partial E}{\partial e} \frac{\partial e}{\partial \hat{y}} \frac{\partial \hat{y}}{\partial (\bar{\mu}_i f_i)} \frac{\partial (\bar{\mu}_i f_i)}{\partial \bar{\mu}_i} \frac{\partial \bar{\mu}_i}{\partial \mu_i} \frac{\partial \mu_i}{\partial \mu_.(.)} \frac{\partial \mu_.(.)}{\partial a_i}$$

where α is the learning rate and $f_i = y_i$.

Following the same process one can show that

$$\Delta b_i = -\alpha \frac{\partial E}{\partial e} \frac{\partial e}{\partial \hat{y}} \frac{\partial \hat{y}}{\partial (\bar{\mu}_i f_i)} \frac{\partial (\bar{\mu}_i f_i)}{\partial \bar{\mu}_i} \frac{\partial \bar{\mu}_i}{\partial \mu_i} \frac{\partial \mu_i}{\partial \mu_.(.)} \frac{\partial \mu_.(.)}{\partial b_i}$$

and

$$\Delta c_i = -\alpha \frac{\partial E}{\partial e} \frac{\partial e}{\partial \hat{y}} \frac{\partial \hat{y}}{\partial (\bar{\mu}_i f_i)} \frac{\partial (\bar{\mu}_i f_i)}{\partial \bar{\mu}_i} \frac{\partial \bar{\mu}_i}{\partial \mu_i} \frac{\partial \mu_i}{\partial \mu_.(.)} \frac{\partial \mu_.(.)}{\partial c_i}$$

where $\mu_.(.)$ depends on the value of i.

We know that,

$E = \frac{1}{2} e^2$ and so $\frac{\partial E}{\partial e} = e$

Again, $\frac{1}{2} e^2 = \frac{1}{2} (y - \hat{y})^2$

Taking partial derivative with respect to y on both sides we get

$$e \frac{\partial e}{\partial \hat{y}} = -(y - \hat{y})$$

Now, we have

$$y = \hat{y} + e$$
$$= \bar{\mu}_1[\beta_{10} + \beta_{11}x_1 + \beta_{12}x_2] + \bar{\mu}_2[\beta_{20} + \beta_{21}x_1 + \beta_{22}x_2] + \bar{\mu}_3[\beta_{30} + \beta_{31}x_1 + \beta_{32}x_2] +$$
$$\bar{\mu}_4[\beta_{40} + \beta_{41}x_1 + \beta_{42}x_2] + e$$
$$= \sum_{i=1}^{4} \bar{\mu}_i f_i + e$$

This shows that

$$\frac{\partial \hat{y}}{\partial(\bar{\mu}_i f_i)} \frac{\partial(\bar{\mu}_i f_i)}{\partial \bar{\mu}_i} = f_i$$

Further, we have

$$\bar{\mu}_i = \frac{\mu_i}{\sum\limits_{i=1}^{4} \mu_i}$$

Thus,

$$\frac{\partial \bar{\mu}_i}{\partial \mu_i} = \frac{\left(\sum\limits_{i=1}^{4} \mu_i\right) - \mu_i}{\left(\sum\limits_{i=1}^{4} \mu_i\right)^2} = \frac{\bar{\mu}_i(1 - \bar{\mu}_i)}{\mu_i}$$

Now, for $i = 1$,

$$\mu_1 = \mu_L(x_1)\mu_L(x_2)$$
$$\therefore \frac{\partial \mu_1}{\partial \mu_L(x_1)} = \mu_L(x_2) = \frac{\mu_1}{\mu_L(x_1)}$$

and

$$\frac{\partial \mu_L(x_1)}{\partial c_1} = \mu_L^2(x_1)\left(\frac{2b_1}{a_1}\right)\left(\frac{x_1 - c_1}{a_1}\right)^{2b_1 - 1}$$

Thus,

$$\Delta c_1 = \alpha(y - \hat{y})f_1\bar{\mu}_1(1 - \bar{\mu}_1)\mu_L(x_1)\left(\frac{2b_1}{a_1}\right)\left(\frac{x_1 - c_1}{a_1}\right)^{2b_1 - 1} \tag{7.4}$$

Similarly, it can be shown that

$$\frac{\partial \mu_L(x_1)}{\partial b_1} = -\frac{2\left(\frac{x_1 - c_1}{a_1}\right)^{2b_1} \log\left(\frac{x_1 - c_1}{a_1}\right)}{\left(1 + \left(\frac{x_1 - c_1}{a_1}\right)^{2b_1}\right)^2} = -2\mu_L^2(x_1)\left(\frac{x_1 - c_1}{a_1}\right)^{2b_1} \log\left(\frac{x_1 - c_1}{a_1}\right)$$

and

$$\frac{\partial \mu_L(x_1)}{\partial a_1} = 2 \frac{b_1}{a_1} \mu_L^2(x_1) \left(\frac{x_1 - c_1}{a_1} \right)^{2b_1}$$

Hence,

$$\Delta b_1 = -2\alpha(y - \hat{y})f_1\bar{\mu}_1(1 - \bar{\mu}_1)\mu_L(x_1)\left(\frac{2b_1}{a_1}\right)\left(\frac{x_1 - c_1}{a_1}\right)^{2b_1}\log\left(\frac{x_1 - c_1}{a_1}\right) \tag{7.5}$$

and

$$\Delta a_1 = 2\alpha(y - \hat{y})f_1\bar{\mu}_1(1 - \bar{\mu}_1)\mu_L(x_1)\left(\frac{b_1}{a_1}\right)\left(\frac{x_1 - c_1}{a_1}\right)^{2b_1} \tag{7.6}$$

Similarly, the small changes of the remaining parameters can be obtained for $i = 2, 3, 4$.

7.2.1 Step-by-step working principle of ANFIS

Let us consider the following testing data of Table 7.1 to explain the working principle of ANFIS. Table 7.2 shows the initial values of the parameters for bell-shaped membership functions.

Considering the 1st training set we have, $x_1 = 4.4$, $x_2 = 13.8$, and $y = 15.3$. The input-output relationships of the different layers are found to be as follows:

Layer 1:

$$y_1^{(1)} = x_1^{(1)} = 4.4$$
$$y_2^{(1)} = x_2^{(1)} = 13.8$$

Layer 2:

$$x_1^{(2)} = 4.4$$
$$x_2^{(2)} = 4.4$$
$$x_3^{(2)} = 13.8$$
$$x_4^{(2)} = 13.8$$

$$y_1^{(2)} = \frac{1}{1 + \left|\frac{4.4-3}{0.5}\right|^{2\times1}} = \mu_L(x_1) = 0.1131$$

$$y_2^{(2)} = \frac{1}{1 + \left|\frac{4.4-8}{0.5}\right|^{2\times1}} = \mu_H(x_1) = 0.0189$$

$$y_3^{(2)} = \frac{1}{1 + \left|\frac{13.8-8}{4}\right|^{2\times2}} = \mu_L(x_2) = 0.1845$$

$$y_4^{(2)} = \frac{1}{1 + \left|\frac{13.8-20}{4}\right|^{2\times2}} = \mu_H(x_2) = 0.1477$$

TABLE 7.1 Training data.

x_1	x_2	y
4.4	13.8	15.3
3.7	18.4	13.98
4.3	9.8	16.24
3.8	8.4	14.49
4.5	8.4	14.6
4.5	16.6	15.15
4.4	8.7	16.28
3.8	15.5	14.15
4.7	9.0	16.39
3.9	11.8	14.38
3.8	13.1	15.45
3.1	11.3	15.62
4.7	11.7	15.25
4.2	10.8	15.26
4.1	9.2	17.64

TABLE 7.2 Initial values of the parameters.

Input	Fuzzy subset	Parameters		
		a_i	b_i	c_i
x_1	Low	0.5	1	3
	High	0.5	1	8
x_2	Low	4	2	8
	High	4	2	20

Layer 3:

$$x_1^{(3)} = [0.1131, 0.1845]$$
$$x_2^{(3)} = [0.1131, 0.1477]$$
$$x_3^{(3)} = [0.0189, 0.1845]$$
$$x_4^{(3)} = [0.0189, 0.1477]$$

$$y_1^{(3)} = y_1^{(2)} y_3^{(2)} = \mu_L(x_1)\mu_L(x_2) = \mu_1 = 0.1131 \times 0.1845 = 0.0209$$
$$y_2^{(3)} = y_1^{(2)} y_4^{(2)} = \mu_L(x_1)\mu_H(x_2) = \mu_2 = 0.1131 \times 0.1477 = 0.0167$$
$$y_3^{(3)} = y_2^{(2)} y_3^{(2)} = \mu_H(x_1)\mu_L(x_2) = \mu_3 = 0.0189 \times 0.1845 = 0.0035$$
$$y_4^{(3)} = y_2^{(2)} y_4^{(2)} = \mu_H(x_1)\mu_H(x_2) = \mu_4 = 0.0189 \times 0.1477 = 0.0028$$

Layer 4:

$$x_1^{(4)} = x_2^{(4)} = x_3^{(4)} = x_4^{(4)} = [\mu_1, \mu_2, \mu_3, \mu_4] = [0.0209, 0.0167, 0.0035, 0.0028]$$

The corresponding outputs are

$$y_1^{(4)} = \bar{\mu}_1 = \frac{\mu_1}{\mu_1 + \mu_2 + \mu_3 + \mu_4} = \frac{0.0209}{0.0209 + 0.0167 + 0.0035 + 0.0028} = 0.4758$$

$$y_2^{(4)} = \bar{\mu}_2 = \frac{\mu_2}{\mu_1 + \mu_2 + \mu_3 + \mu_4} = \frac{0.0167}{0.0209 + 0.0167 + 0.0035 + 0.0028} = 0.3809$$

$$y_3^{(4)} = \bar{\mu}_3 = \frac{\mu_3}{\mu_1 + \mu_2 + \mu_3 + \mu_4} = \frac{0.0035}{0.0209 + 0.0167 + 0.0035 + 0.0028} = 0.0796$$

$$y_4^{(4)} = \bar{\mu}_4 = \frac{\mu_4}{\mu_1 + \mu_2 + \mu_3 + \mu_4} = \frac{0.0028}{0.0209 + 0.0167 + 0.0035 + 0.0028} = 0.0637$$

Layer 5:

$$x_1^{(5)} = \bar{\mu}_1 = 0.4758$$
$$x_2^{(5)} = \bar{\mu}_2 = 0.3809$$
$$x_3^{(5)} = \bar{\mu}_3 = 0.0796$$
$$x_4^{(5)} = \bar{\mu}_4 = 0.0637$$

$$y_1^{(5)} = \bar{\mu}_1 y_1 = \bar{\mu}_1[\beta_{10} + \beta_{11}x_1 + \beta_{12}x_2] = 0.4758[\beta_{10} + 4.4\beta_{11} + 13.8\beta_{12}]$$
$$y_2^{(5)} = \bar{\mu}_2 y_2 = \bar{\mu}_2[\beta_{20} + \beta_{21}x_1 + \beta_{22}x_2] = 0.3809[\beta_{20} + 4.4\beta_{21} + 13.8\beta_{22}]$$
$$y_3^{(5)} = \bar{\mu}_3 y_3 = \bar{\mu}_3[\beta_{30} + \beta_{31}x_1 + \beta_{32}x_2] = 0.0796[\beta_{30} + 4.4\beta_{31} + 13.8\beta_{32}]$$
$$y_4^{(5)} = \bar{\mu}_4 y_4 = \bar{\mu}_4[\beta_{40} + \beta_{41}x_1 + \beta_{42}x_2] = 0.0637[\beta_{40} + 4.4\beta_{41} + 13.8\beta_{42}]$$

Layer 6:

$$x_1^{(6)} = [\bar{\mu}_1 y_1, \bar{\mu}_2 y_2, \bar{\mu}_3 y_3, \bar{\mu}_4 y_4]$$

$$\hat{y} = y_1^{(6)} = \bar{\mu}_1 y_1 + \bar{\mu}_2 y_2 + \bar{\mu}_3 y_3 + \bar{\mu}_4 y_4$$

$$= 0.4758[\beta_{10} + 4.4\beta_{11} + 13.8\beta_{12}] + 0.3809[\beta_{20} + 4.4\beta_{21} + 13.8\beta_{22}] +$$

$$0.0796[\beta_{30} + 4.4\beta_{31} + 13.8\beta_{32}] + 0.0637[\beta_{40} + 4.4\beta_{41} + 13.8\beta_{42}]$$

$$= 0.4758\beta_{10} + 2.0936\beta_{11} + 6.5663\beta_{12} + 0.3809\beta_{20} + 1.6758\beta_{21} + 5.2559\beta_{22} +$$

$$0.0796\beta_{30} + 0.3503\beta_{31} + 1.0985\beta_{32} + 0.0637\beta_{40} + 0.2804\beta_{41} + 0.8793\beta_{42}$$

Thus, it becomes

$$y(1) = 0.4758\beta_{10} + 2.0936\beta_{11} + 6.5663\beta_{12} + 0.3809\beta_{20} + 1.6758\beta_{21} +$$
$$5.2559\beta_{22} + 0.0796\beta_{30} + 0.3503\beta_{31} + 1.0985\beta_{32} + 0.0637\beta_{40} +$$
$$0.2804\beta_{41} + 0.8793\beta_{42} + e(1)$$

Similarly, we can obtain

$$y(2) = 0.0207\beta_{10} + 0.0765\beta_{11} + 0.3804\beta_{12} + 0.9413\beta_{20} + 3.483\beta_{21} +$$
$$17.3206\beta_{22} + 0.0008\beta_{30} + 0.003\beta_{31} + 0.015\beta_{32} + 0.0372\beta_{40} +$$
$$0.1375\beta_{41} + 0.684\beta_{42} + e(2)$$

$$y(3) = 0.8572\beta_{10} + 3.686\beta_{11} + 8.4007\beta_{12} + 0.0206\beta_{20} + 0.0887\beta_{21} +$$
$$0.202\beta_{22} + 0.1193\beta_{30} + 0.513\beta_{31} + 1.1691\beta_{32} + 0.0029\beta_{40} +$$
$$0.0123\beta_{41} + 0.0281\beta_{42} + e(3)$$

$$y(4) = 0.9395\beta_{10} + 3.5701\beta_{11} + 7.8919\beta_{12} + 0.0131\beta_{20} + 0.0498\beta_{21} +$$
$$0.11\beta_{22} + 0.0467\beta_{30} + 0.1776\beta_{31} + 0.3926\beta_{32} + 0.0007\beta_{40} +$$
$$0.0025\beta_{41} + 0.0055\beta_{42} + e(4)$$

$$y(5) = 0.8219\beta_{10} + 3.6984\beta_{11} + 6.9037\beta_{12} + 0.0115\beta_{20} + 0.0516\beta_{21} +$$
$$0.0963\beta_{22} + 0.1644\beta_{30} + 0.7397\beta_{31} + 1.3807\beta_{32} + 0.0023\beta_{40} +$$
$$0.0103\beta_{41} + 0.0193\beta_{42} + e(5)$$

$$y(6) = 0.0531\beta_{10} + 0.2389\beta_{11} + 0.8813\beta_{12} + 0.7802\beta_{20} + 3.5111\beta_{21} +$$
$$12.952\beta_{22} + 0.0106\beta_{30} + 0.0478\beta_{31} + 0.1763\beta_{32} + 0.156\beta_{40} +$$
$$0.7022\beta_{41} + 2.5904\beta_{42} + e(6)$$

$$y(7) = 0.8436\beta_{10} + 3.712\beta_{11} + 7.3395\beta_{12} + 0.0131\beta_{20} + 0.0574\beta_{21} +$$
$$0.1136\beta_{22} + 0.1411\beta_{30} + 0.621\beta_{31} + 1.2279\beta_{32} + 0.0022\beta_{40} +$$
$$0.0096\beta_{41} + 0.019\beta_{42} + e(7)$$

$$y(8) = 0.1553\beta_{10} + 0.5901\beta_{11} + 2.4069\beta_{12} + 0.7973\beta_{20} + 3.0298\beta_{21} +$$
$$12.3586\beta_{22} + 0.0077\beta_{30} + 0.0294\beta_{31} + 0.1197\beta_{32} + 0.0397\beta_{40} +$$
$$0.1507\beta_{41} + 0.6148\beta_{42} + e(8)$$

$$y(9) = 0.7669\beta_{10} + 3.6043\beta_{11} + 6.9019\beta_{12} + 0.0132\beta_{20} + 0.0622\beta_{21} +$$
$$0.1191\beta_{22} + 0.2162\beta_{30} + 1.0159\beta_{31} + 1.9454\beta_{32} + 0.0037\beta_{40} +$$
$$0.0175\beta_{41} + 0.0336\beta_{42} + e(9)$$

$$y(10) = 0.8581\beta_{10} + 3.3465\beta_{11} + 10.1252\beta_{12} + 0.0834\beta_{20} + 0.3254\beta_{21} +$$
$$0.9845\beta_{22} + 0.0533\beta_{30} + 0.2079\beta_{31} + 0.6291\beta_{32} + 0.0052\beta_{40} +$$
$$0.0202\beta_{41} + 0.0612\beta_{42} + e(10)$$

$$y(11) = 0.6955\beta_{10} + 2.6429\beta_{11} + 9.1112\beta_{12} + 0.2571\beta_{20} + 0.977\beta_{21} +$$
$$3.368\beta_{22} + 0.0346\beta_{30} + 0.1315\beta_{31} + 0.4533\beta_{32} + 0.0128\beta_{40} +$$
$$0.0486\beta_{41} + 0.1676\beta_{42} + e(11)$$

$$y(12) = 0.9311\beta_{10} + 2.8865\beta_{11} + 10.5216\beta_{12} + 0.0583\beta_{20} + 0.1807\beta_{21} +$$
$$0.6585\beta_{22} + 0.01\beta_{30} + 0.0309\beta_{31} + 0.1128\beta_{32} + 0.0006\beta_{40} +$$
$$0.0019\beta_{41} + 0.0071\beta_{42} + e(12)$$

$$y(13) = 0.7166\beta_{10} + 3.368\beta_{11} + 8.3841\beta_{12} + 0.0635\beta_{20} + 0.2986\beta_{21} +$$
$$0.7433\beta_{22} + 0.202\beta_{30} + 0.9493\beta_{31} + 2.3632\beta_{32} + 0.0179\beta_{40} +$$
$$0.0842\beta_{41} + 0.2095\beta_{42} + e(13)$$

$$y(14) = 0.86\beta_{10} + 3.6121\beta_{11} + 9.2883\beta_{12} + 0.0368\beta_{20} + 0.1545\beta_{21} +$$
$$0.3974\beta_{22} + 0.0989\beta_{30} + 0.4156\beta_{31} + 1.0686\beta_{32} + 0.0042\beta_{40} +$$
$$0.0178\beta_{41} + 0.0457\beta_{42} + e(14)$$

$$y(15) = 0.897\beta_{10} + 3.6777\beta_{11} + 8.2525\beta_{12} + 0.0167\beta_{20} + 0.0685\beta_{21} +$$
$$0.1537\beta_{22} + 0.0847\beta_{30} + 0.3473\beta_{31} + 0.7793\beta_{32} + 0.0016\beta_{40} +$$
$$0.0065\beta_{41} + 0.0145\beta_{42} + e(15)$$

Using the matrix notation, we have

$$Y = \begin{bmatrix} 15.3 \\ 13.98 \\ 16.24 \\ 14.49 \\ 14.6 \\ 15.15 \\ 16.28 \\ 14.15 \\ 16.39 \\ 14.38 \\ 15.45 \\ 15.62 \\ 15.25 \\ 15.26 \\ 17.64 \end{bmatrix}$$

$$X = \begin{bmatrix}
0.4758 & 2.0936 & 6.5663 & 0.3809 & 1.6758 & 5.2559 & 0.0796 & 0.3503 & 1.0985 & 0.0637 & 0.2804 & 0.8793 \\
0.0207 & 0.0765 & 0.3804 & 0.9413 & 3.483 & 17.3206 & 0.0008 & 0.003 & 0.015 & 0.0372 & 0.1375 & 0.684 \\
0.8572 & 3.686 & 8.4007 & 0.0206 & 0.0887 & 0.202 & 0.1193 & 0.513 & 1.1691 & 0.0029 & 0.0123 & 0.0281 \\
0.9395 & 3.5701 & 7.8919 & 0.0131 & 0.0498 & 0.11 & 0.0467 & 0.1776 & 0.3926 & 0.0007 & 0.0025 & 0.0055 \\
0.8219 & 3.6984 & 6.9037 & 0.0115 & 0.0516 & 0.0963 & 0.1644 & 0.7397 & 1.3807 & 0.0023 & 0.0103 & 0.0193 \\
0.0531 & 0.2389 & 0.8813 & 0.7802 & 3.5111 & 12.952 & 0.0106 & 0.0478 & 0.1763 & 0.156 & 0.7022 & 2.5904 \\
0.8436 & 3.712 & 7.3395 & 0.0131 & 0.0574 & 0.1136 & 0.1411 & 0.621 & 1.2279 & 0.0022 & 0.0096 & 0.019 \\
0.1553 & 0.5901 & 2.4069 & 0.7973 & 3.0298 & 12.3586 & 0.0077 & 0.0294 & 0.1197 & 0.0397 & 0.1507 & 0.6148 \\
0.7669 & 3.6043 & 6.9019 & 0.0132 & 0.0622 & 0.1191 & 0.2162 & 1.0159 & 1.9454 & 0.0037 & 0.0175 & 0.0336 \\
0.8581 & 3.3465 & 10.1252 & 0.0834 & 0.3254 & 0.9845 & 0.0533 & 0.2079 & 0.6291 & 0.0052 & 0.0202 & 0.0612 \\
0.6955 & 2.6429 & 9.1112 & 0.2571 & 0.977 & 3.368 & 0.0346 & 0.1315 & 0.4533 & 0.0128 & 0.0486 & 0.1676 \\
0.9311 & 2.8865 & 10.5216 & 0.0583 & 0.1807 & 0.6585 & 0.01 & 0.0309 & 0.1128 & 0.0006 & 0.0019 & 0.0071 \\
0.7166 & 3.368 & 8.3841 & 0.0635 & 0.2986 & 0.7433 & 0.202 & 0.9493 & 2.3632 & 0.0179 & 0.0842 & 0.2095 \\
0.86 & 3.6121 & 9.2883 & 0.0368 & 0.1545 & 0.3974 & 0.0989 & 0.4156 & 1.0686 & 0.0042 & 0.0178 & 0.0457 \\
0.897 & 3.6777 & 8.2525 & 0.0167 & 0.0685 & 0.1537 & 0.0847 & 0.3473 & 0.7793 & 0.0016 & 0.0065 & 0.0145
\end{bmatrix}$$

Thus, the values of the coefficients become

$$B = (X^T X)^{-1} X^T Y = \begin{bmatrix} -176.59 \\ 61.534 \\ -1.7331 \\ 4201.2 \\ -1219.9 \\ -11.37 \\ -1864.4 \\ 313.01 \\ 14.617 \\ 39609 \\ -7452.7 \\ 84.405 \end{bmatrix}$$

Hence, the predicted output matrix is obtained as

$$\hat{Y} = \begin{bmatrix} 15.288 \\ 14.025 \\ 16.361 \\ 15.167 \\ 15.414 \\ 15.157 \\ 15.882 \\ 14.008 \\ 16.126 \\ 13.927 \\ 15.828 \\ 15.545 \\ 15.203 \\ 16.091 \\ 16.158 \end{bmatrix}$$

The actual and predicted values of outputs and the corresponding error terms are shown in Table 7.3.

From Table 7.3, the error sum of square is estimated as

$$E = \frac{1}{2} \sum_{i=1}^{15} e_i^2 = 2.3144$$

TABLE 7.3 Actual and predicted outputs with errors.

y	$\hat{y}$	e
15.3	15.288	0.012
13.98	14.025	−0.045
16.24	16.361	−0.121
14.49	15.167	−0.677
14.6	15.414	−0.814
15.15	15.157	−0.007
16.28	15.882	0.398
14.15	14.008	0.142
16.39	16.126	0.264
14.38	13.927	0.453
15.45	15.828	−0.378
15.62	15.545	0.075
15.25	15.203	0.047
15.26	16.091	−0.831
17.64	16.158	1.482

In this example, the total number of linear parameters (N_1) and nonlinear parameters (N_2) can be calculated using Eqs. (7.2) and (7.3) as follows:

$$N_1 = (n + 1) \times p^n$$
$$= (2 + 1) \times 2^2$$
$$= 12$$

$$N_2 = n \times p \times F$$
$$= 2 \times 2 \times 3$$
$$= 12$$

In the given example, assuming learning rate of 0.9, the first input data point and its linguistic term at "low" level gives

$$\alpha = 0.9$$

$$(y - \hat{y}) = 0.012$$

$$f_1 \bar{\mu}_1 = 0.4758[\beta_{10} + 4.4\beta_{11} + 13.8\beta_{12}]$$

where $\beta_{10} = -176.59$, $\beta_{11} = 61.534$, $\beta_{12} = -1.7331$,

$$\mu_L(x_1) = 0.1131$$

$$x_1 = 4.4$$

Hence, using Eqs. (7.4)–(7.6), we have

$$\Delta c_1 = 0.2397$$

$$\Delta b_1 = -0.3455$$

$$\Delta a_1 = 0.6711$$

In a similar way, other values of the parameters a_i, b_i, and c_i will be upgraded for the 1st training set. The calculation for 2nd training set will resume with the upgraded values of the parameters a_i, b_i, and c_i of the 1st training set. This process will continue until the last training set and with this first epoch will be completed. The similar steps will be followed in the subsequent epochs until the stopping criteria are reached.

7.2.2 Application of ANFIS in prediction of yarn irregularity

Cotton fibers with a different combination of micronaire value (MIC) and short fiber content (SFC) were spun into yarns of three different English counts (10^s, 22^s, and 30^s Ne) in rotor spinning system. Table 7.4 shows a total of 100 data points comprising MIC, SFC, yarn English count (Ne), and yarn unevenness

TABLE 7.4 Dataset of cotton fiber and yarn.

MIC (µg/in.)	SFC (%)	Yarn count (Ne)	Yarn CV_m%
3.8	8.4	29.7	13.66
4.6	10.3	29.8	14.35
3.7	8.9	10	10.26
4.7	11.7	21.7	12.18
4.6	9.2	22.1	12.34
3.8	13.1	21.7	11.9
4.3	9.8	21.9	11.94
4.3	5.6	10.2	10.1
3.5	11.9	10.1	10.1
4.5	16.6	9.9	11.12

TABLE 7.4 Dataset of cotton fiber and yarn—cont'd

MIC (µg/in.)	SFC (%)	Yarn count (Ne)	Yarn CV$_m$%
4.5	7.2	30.3	13.89
3.8	8.4	10	10.11
4.1	9.2	10	10.22
3.1	11.3	9.9	10.3
5	6.8	10.1	10.27
3.8	15.5	9.9	10.78
4.7	9	21.9	12.22
4.4	8.7	30.2	14.36
3.7	18.4	10	10.4
3.7	8.9	22.6	11.66
4.4	8.7	10.1	10.5
3.7	6.8	29.9	13.46
3.1	11.3	21.8	11.78
3.7	6.8	10	10.18
4.9	6.8	30.1	14.21
3.8	6.8	10	9.78
3.5	11.9	21.8	11.94
4.9	6.8	10.1	10.62
4.6	9.2	30.2	13.97
3.8	13.1	30.5	13.78
3.1	11.3	29.8	13.78
4.5	7.2	10.2	10.24
4.5	16.6	29.8	14.44
4.4	8.7	21.5	12.38
4.6	5.6	22.2	12.1
3.8	6.8	22.1	11.34
4.8	7.2	10.1	10.4
4.2	10.8	10	10.66
4.3	10.2	10.1	10.54
3.9	11.8	21.9	12.45

Continued

TABLE 7.4 Dataset of cotton fiber and yarn—cont'd

MIC (µg/in.)	SFC (%)	Yarn count (Ne)	Yarn CV$_m$%
4.6	10.3	10.1	11.2
4.8	7.5	10	10.61
4.2	10.8	30	14.21
4.6	7.7	10	10.07
4.1	9.2	29.6	14.16
4.5	16.6	21.8	12.47
3.8	13.1	10.1	10.34
4.5	7.2	22.3	12.07
4.7	11.7	10.2	10.29
4.2	10.9	21.5	12.07
3.8	8.7	22.2	11.77
4.3	9.8	30	13.85
3.8	8.7	29.9	13.54
4.4	13.8	29.8	13.77
4.4	8.1	30.1	13.49
3.8	6.8	30.4	13.23
3.8	8.7	10.1	10.02
4	11.2	30	14.11
4	11.2	21.8	12.05
4.8	7.2	22	12.18
4.8	7.2	30.6	14.22
3.9	11.8	30	14.31
4.6	7.7	30.5	13.94
3.7	18.4	30.3	13.78
4.6	5.6	10.1	10.1
4.7	9	10.1	10.32
4.6	8.4	10	9.94
4.6	9.2	10.1	10.7
3.8	15.5	29.6	14.34
4.6	8.4	30.6	13.76

TABLE 7.4 Dataset of cotton fiber and yarn—cont'd

MIC (μg/in.)	SFC (%)	Yarn count (Ne)	Yarn CV$_m$%
3.9	11.8	10	10.71
4.6	8.4	22	11.66
4.8	7.5	22	12.33
4.5	9.7	30.2	14.42
4.4	8.1	22	11.78
3.7	8.9	30.4	13.55
3.7	6.8	21.9	11.88
4.4	13.8	10.2	10.3
4.4	8.1	10.1	10.04
4.6	5.6	30.2	13.92
4.4	13.8	21.6	12.03
4.5	8.4	10.1	10.51
4.3	5.6	30.4	14.04
4.8	7.5	30	14.19
4.1	9.2	21.9	12.06
4.5	9.7	22.1	12.16
4.2	10.9	10	10.16
4.6	7.7	22.4	11.97
4.5	8.4	21.8	12.34
4.5	9.7	10	10.58
4.6	10.3	22.2	12.69
3.7	18.4	21.7	11.9
4.9	6.8	22.2	12.38
4.7	9	29.9	14.45
4.5	8.4	29.6	14.38
3.8	15.5	21.7	12.52
4.3	10.2	29.7	13.68
4.2	10.8	21.8	12.23
4.7	11.7	30.4	14.31
4.2	10.9	30.1	14.06

(CV_m%). Three inputs such as MIC, SFC, and yarn count (Ne) are considered for the prediction of yarn unevenness using ANFIS (Ghosh, 2010).

A sequence of arithmetic progress with a common difference of 10, such as 10th, 20th, 30th, ..., 100th data points, is used as the testing dataset, and remaining 90 data points are used as the training dataset for the ANFIS model. A bell-shaped membership function is selected in this model and for each input two membership functions are used. As there are three inputs, the total number of fuzzy rules becomes $2^3 = 8$. Using Eq. (7.2), the total number of linear parameters (N_1) is calculated to be $(3 + 1) \times 2^3 = 32$. As the bell-shaped membership function employs three parameters, from Eq. (7.3) we have the total number of nonlinear parameters is $3 \times 2 \times 3 = 18$. The training parameters such as the maximum number of iterations, error goal, initial step size, step size decrease rate, and step size increase rate are set to 40,000, 10^{-5}, 0.01, 0.9, and 1.1, respectively.

The mean absolute percentage accuracy between the actual and predicted values for training and testing sets is estimated as 98.85% and 98.57%, respectively. Table 7.5 shows the prediction performance for the testing dataset. The learning accuracy on the training set was expectedly higher than the predictive accuracy on the test set due to the fact that the latter is performed on the unseen dataset. A MATLAB® coding was used to execute the computational work. Figs. 7.2–7.4 show the membership functions before and after the training for the input parameters, namely micronaire value, short fiber content, and yarn count, respectively.

TABLE 7.5 Testing accuracies.

Data serial number	Actual CV_m (%)	Predicted CV_m (%)	Prediction accuracy (%)
10	11.12	10.75	96.67
20	11.66	12.23	95.10
30	13.78	13.72	99.55
40	12.45	12.37	99.36
50	12.07	12.11	99.70
60	12.18	12.18	99.97
70	13.76	14.20	96.78
80	13.92	13.98	99.55
90	10.58	10.53	99.52
100	14.06	14.13	99.53
Mean accuracy (%) of prediction on testing set			98.57

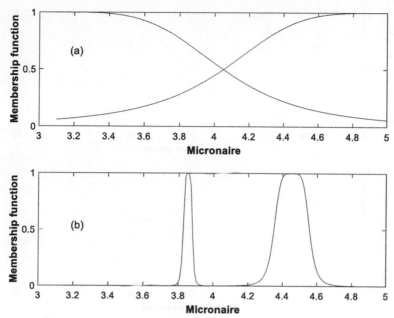

FIG. 7.2 Initial (A) and final (B) membership functions of micronaire value.

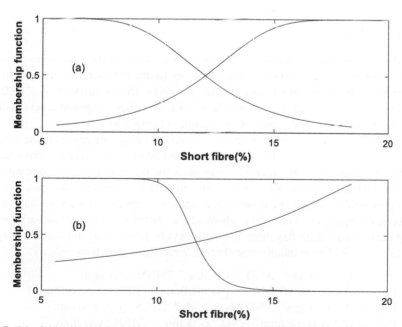

FIG. 7.3 Initial (A) and final (B) membership functions of short fiber content (%).

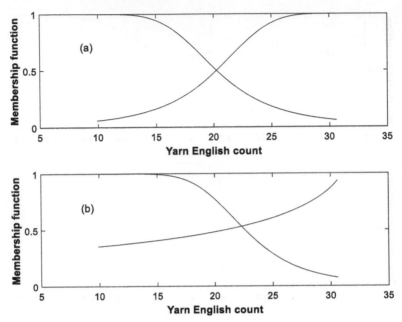

FIG. 7.4 Initial (A) and final (B) membership functions of yarn count.

7.3 GFES

In this system the binary-coded GA is used to improve the performance of Mamdani based fuzzy expert system (FES) by tuning the membership function distributions of the input and output variables (Shimojima et al., 1995; Damousis et al., 2002). The principle of GFES has been explained here with the help of the following numerical example (Table 7.6).

In this example, we have two inputs (x_1 and x_2) and one output (y) in the crisp form. At first, the fuzzification of the inputs and output is done. Here, triangular form of membership function has been used for the purpose of fuzzification. Three linguistic fuzzy subsets namely low (L), medium (M), and high (H) are chosen for each of the two inputs and one output in such a way that they are equally spaced and cover the whole space. As there are two input variables and each one of them has three linguistic levels, hence total number of fuzzy rules is $3^2 = 9$. The manually constructed 9 rules are shown as follows:

Rule 1: IF "x_1 is low" AND "x_2 is low," THEN "y is high."
Rule 2: IF "x_1 is low" AND "x_2 is medium," THEN "y is medium."
Rule 3: IF "x_1 is low" AND "x_2 is high," THEN "y is medium."
Rule 4: IF "x_1 is medium" AND "x_2 is low," THEN "y is high."
Rule 5: IF "x_1 is medium" AND "x_2 is medium," THEN "y is medium."

TABLE 7.6 Training cases.

x_1	x_2	y
4.4	13.8	15.3
3.7	18.4	13.98
4.3	9.8	16.24
3.8	8.4	14.49
4.5	8.4	14.6
4.5	16.6	15.15
4.4	8.7	16.28
3.8	15.5	14.15
4.7	9.0	16.39
3.9	11.8	14.38
3.8	13.1	15.45
3.1	11.3	15.62
4.7	11.7	15.25
4.2	10.8	15.26
4.1	9.2	17.64

Rule 6: IF "x_1 is medium" AND "x_2 is high," THEN "y is low."
Rule 7: IF "x_1 is high" AND "x_2 is low," THEN "y is medium."
Rule 8: IF "x_1 is high" AND "x_2 is medium," THEN "y is low."
Rule 9: IF "x_1 is high" AND "x_2 is high," THEN "y is low."

These fuzzy rules are depicted in Table 7.7.

An initial population of the binary coded genetic algorithm (GA) of population size N is created at random. The population size (N) can be large, but for the purpose of simplification, let us suppose that the initial population of only 6 strings is created as random:

String-1 (S1): 101001011101000000110010100011110
String-2 (S2): 100010110010101000011000101000001
String-3 (S3): 011110100110011100110111000001101
String-4 (S4): 100110110111101000000110100110100
String-5 (S5): 001001000110001010001001111001111
String-6 (S6): 011000110111110011000111010110100

TABLE 7.7 Manually constructed fuzzy rules.

		x_2		
		L	M	H
x_1	L	H	M	M
	M	H	M	L
	H	M	L	L

Each of the above strings is composed of four substrings, of which the first three substrings represent the distribution parameters $(b_1, b_2,$ and $b_3)$ of triangular membership functions for two inputs and one output, and the fourth substring represents the fuzzy rule base. More specifically, the values of b_1, b_2, and b_3 determine the half-width of the base of the triangles for the linguistic subsets of two inputs and one output, respectively, as depicted in Fig. 7.5. For each string of the population, starting from the leftmost bit position, 8 bits are assigned to indicate each of the b values (i.e., b_1, b_2, b_3) and the next 9 bits represent the fuzzy rule base (Table 7.8). In the GFES, these b values are optimized to get better-suited membership functions that increase the prediction accuracy of the model by minimizing the mean absolute deviation of actual and predicted output values. Hence, the objective function of the GFES becomes:

$$\text{Minimize } f(b_1, b_2, b_3) = \bar{d}$$

$$\text{for } b_{i(\min)} \leq b_i \leq b_{i(\max)} \text{ and } i = 1, 2, 3$$

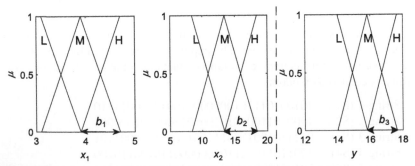

FIG. 7.5   Distribution of membership functions for inputs and output variables.

TABLE 7.8 Initial population of GA strings.

	b_1						b_2						b_3								Rule selection						
S1	1	0	1	0	0	1	0	1	1	0	1	1	0	0	0	1	0	0	1	0	0	1	1	1	1	1	0
S2	1	0	0	0	1	0	1	0	0	1	0	0	0	0	1	0	1	0	0	1	1	0	1	1	0	0	1
S3	0	1	0	1	0	1	0	1	1	0	0	1	1	0	0	1	1	0	1	1	0	0	0	0	1	0	0
S4	1	0	0	1	0	1	0	1	1	1	0	1	0	0	0	0	0	1	0	0	1	1	1	1	1	0	1
S5	0	0	0	0	1	0	0	0	0	0	1	0	1	0	0	0	1	0	1	0	0	1	1	0	0	1	0
S6	0	1	1	0	0	0	1	0	0	1	1	1	1	1	0	0	0	1	1	1	0	1	1	1	1	0	0

where $\bar{d}$ is the mean absolute deviation of actual and predicted output values, $b_{i(\min)}$ and $b_{i(\max)}$ are the lower and upper bounds, respectively, for b_i. In this example, b values are assumed to vary in the ranges given as follows:

$$0.6 \leq b_1 \leq 1$$
$$3 \leq b_2 \leq 7$$
$$1.6 \leq b_3 \leq 2$$

As GA is a maximization procedure, the objective function is modified to the following fitness function

$$F(b_1, b_2, b_3) = \frac{1}{1 + \{f(b_1, b_2, b_3)\}^2}$$

This fitness function is maximized using the binary-coded GA for obtaining the optimum b values. Knowing $b_{i(\min)}$, $b_{i(\max)}$ and the corresponding decoded value of the binary substring, the real value of b_i can be determined using the following linear mapping rule

$$b_i = b_{i(\min)} + \frac{b_{i(\max)} - b_{i(\min)}}{2^{l_i} - 1} \times D_i$$

where l_i is the substring length, D_i is the decoded value corresponding to b_i. For example, in the case of string-1, the first substring (10100101) is decoded as $1 \times 2^7 + 0 \times 2^6 + 1 \times 2^5 + 0 \times 2^4 + 0 \times 2^3 + 1 \times 2^2 + 0 \times 2^1 + 1 \times 2^0 = 165$. Thus, the corresponding real value of b_1 is estimated as follows:

$$b_1 = b_{1(\min)} + \frac{b_{1(\max)} - b_{1(\min)}}{2^{l_1} - 1} \times D_1$$
$$= 0.6 + \frac{1 - 0.6}{2^8 - 1} \times 165$$
$$= 0.6 + \frac{0.4}{256 - 1} \times 165$$
$$= 0.8588$$

The second substring (11010000) of the string-1 can be decoded as $1 \times 2^7 + 1 \times 2^6 + 0 \times 2^5 + 1 \times 2^4 + 0 \times 2^3 + 0 \times 2^2 + 0 \times 2^1 + 0 \times 2^0 = 208$. So, the real value of b_2 is determined as

$$b_2 = b_{2(\min)} + \frac{b_{2(\max)} - b_{2(\min)}}{2^{l_2} - 1} \times D_2$$
$$= 3 + \frac{7 - 3}{2^8 - 1} \times 208$$
$$= 3 + \frac{4}{256 - 1} \times 208$$
$$= 6.2627$$

Similarly, the third substring (00110010) of the string-1 can be decoded as $0 \times 2^7 + 0 \times 2^6 + 1 \times 2^5 + 1 \times 2^4 + 0 \times 2^3 + 0 \times 2^2 + 1 \times 2^1 + 0 \times 2^0 = 50$. Hence, the real value of b_3 is obtained as

$$b_3 = b_{3(\min)} + \frac{b_{3(\max)} - b_{3(\min)}}{2^{l_3} - 1} \times D_3$$

$$= 1.6 + \frac{2 - 1.6}{2^8 - 1} \times 50$$

$$= 1.6 + \frac{0.4}{256 - 1} \times 50$$

$$= 1.6784$$

In a similar way, the real values of b_1, b_2, and b_3 for other strings can be calculated, which are shown in Table 7.9.

The 1 and 0 of the last 9 bits of a GA string denote whether the particular rule from the rule base will be fired or not. In the case of string-1, starting from the left, 1 in the first bit position of the rule base indicates the presence of the first rule, 0 in the second bit position indicates the absence of the second rule, and so on. Accordingly, for string-1, only 5 out of 9 fuzzy rules, viz., rule numbers 1, 5, 6, 7, and 8 will be fired, whereas the remaining 4 rules, i.e., rule numbers 2, 3, 4, and 9 will remain defunct. This is depicted in Table 7.10. Similarly, for string-2, only 3 out of 9 fuzzy rules, viz., rule numbers 1, 3, and 9 will only be fired and the remaining 6 rules will remain defunct. The selected rules to be fired for the other strings are shown in the last column of Table 7.9.

TABLE 7.9 Decoded and real values of strings.

	Decoded values			Real values			
String no.	b_1	b_2	b_3	b_1	b_2	b_3	Selected rules
S1	165	208	50	0.8588	6.2627	1.6784	1, 5, 6, 7, 8
S2	139	42	24	0.8180	3.6588	1.6376	1, 3, 9
S3	122	103	55	0.7914	4.6157	1.6863	6, 7, 9
S4	155	122	6	0.8431	4.9137	1.6094	1, 4, 5, 7
S5	36	98	137	0.6565	4.5373	1.8149	1, 2, 3, 6, 7, 8, 9
S6	99	124	199	0.7553	4.9451	1.9122	2, 4, 5, 7

TABLE 7.10 Selected firing rules for string-1.

		x_2		
		L	M	H
x_1	L	H	–	–
	M	–	M	L
	H	M	L	–

Now, for string-1, using the 1st training data point shown in Table 7.6, the predicted value of output is calculated as follows:

Here both the inputs $x_1 = 4.4$ and $x_2 = 13.8$ may be declared either medium (M) or high (H) and the following 5 rules are being fired from a total of 9 rules.

Rule 1: If x_1 is low AND x_2 is low, THEN y is high.
Rule 5: If x_1 is medium AND x_2 is medium, THEN y is medium.
Rule 6: If x_1 is medium AND x_2 is high, THEN y is low.
Rule 7: If x_1 is high AND x_2 is low, THEN y is medium.
Rule 8: If x_1 is high AND x_2 is medium, THEN y is low.

Using the Mamdani approach with the fuzzy membership functions formed by the corresponding b values ($b_1 = 0.8588$, $b_2 = 6.2627$, $b_3 = 1.6784$) and firing above 5 rules, the fuzzified output corresponding to $x_1 = 4.4$ and $x_2 = 13.8$ can be obtained and the centroid method of defuzzification is used to calculate the crisp value of the output, which is worked out to be 15.45. Fig. 7.6 is an

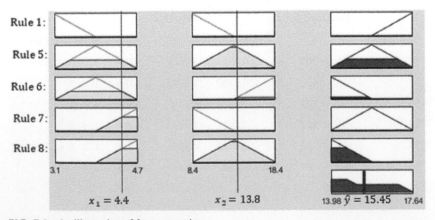

FIG. 7.6 An illustration of fuzzy operation.

TABLE 7.11 Actual and predicted outputs and their deviations.

y_i	$\hat{y}_i$	$y_i - \hat{y}_i$	$d_i = |y_i - \hat{y}_i|$
15.3	15.45	−0.15	0.15
13.98	14.61	−0.63	0.63
16.24	15.72	0.52	0.52
14.49	16.79	−2.30	2.30
14.6	15.81	−1.21	1.21
15.15	15.54	−0.39	0.39
16.28	15.81	0.48	0.48
14.15	15.65	−1.50	1.50
16.39	15.80	0.59	0.59
14.38	15.81	−1.43	1.43
15.45	15.81	−0.36	0.36
15.62	16.91	−1.29	1.29
15.25	15.40	−0.15	0.15
15.26	15.66	−0.40	0.40
17.64	15.75	1.89	1.89

illustration of the fuzzy operation. The absolute value of deviation in prediction (d_1) can be estimated as follows

$$d_1 = |y_1 - \hat{y}_1|$$
$$= |15.3 - 15.45|$$
$$= 0.15$$

Following the similar procedure, the values of absolute deviation in predictions ($d_2, d_3, ..., d_n$) can be determined for the 2nd, 3rd, ..., nth training data points, respectively. Table 7.11 shows the absolute values of deviation in prediction for the training dataset. Thus, the mean of absolute values of deviation for string-1 becomes

$$\bar{d}_1 = \frac{\sum_{i=1}^{n} d_i}{n}$$
$$= \frac{0.15 + 0.63 + ... + 1.89}{15}$$
$$= 0.8855$$

TABLE 7.12 Values of objective and fitness functions for different strings.

String No.	GA-string	Objective function	Fitness function
S1	101001011101000000110010100011110	$f_1 = 0.8855$	$F_1 = 0.5605$
S2	100010110010101000011000101000001	$f_2 = 0.9137$	$F_2 = 0.545$
S3	011110100110011100110111000001101	$f_3 = 0.9137$	$F_3 = 0.545$
S4	100110110111101000000110100110100	$f_4 = 0.8954$	$F_4 = 0.555$
S5	001001000110001010001001111001111	$f_5 = 0.8328$	$F_5 = 0.5905$
S6	011000110111110011000111010110100	$f_6 = 0.8328$	$F_6 = 0.5905$

Consequently, for string-1 the value of objective function turns out to be $f_1 = \overline{d}_1 = 0.8855$, therefore, the value of its fitness function becomes

$$
\begin{aligned}
F_1 &= \frac{1}{1 + f_1^2} \\
&= \frac{1}{1 + 0.8855^2} \\
&= \frac{1}{1 + 0.7841} \\
&= 0.5605
\end{aligned}
$$

The same procedure is repeated to evaluate the fitness values of the other strings of the population. Table 7.12 shows the values of objective function and fitness function for all 6 strings.

The population of GA-strings is then modified using different operators such as reproduction, crossover, and mutation in generation after generation and GA will optimize the performance of fuzzy expert system.

7.3.1 Application of GFES in yarn strength prediction with a comparison to FES

Banerjee et al. (2012) tried to improve the prediction performance of FES modeling of cotton yarn strength by developing a GFES. The data of 36 types of cotton fibers and corresponding yarns of 20s Ne nominal count produced by ring spinning system were collected from the industry. The dataset comprises four fiber properties, bundle strength (FS), upper half mean length (UHML), fineness (FF), and SFC, and corresponding yarn strength is illustrated in Table 7.13.

TABLE 7.13 Fiber properties and corresponding yarn strength.

Sl. no.	FS (cN/tex)	UHML (in.)	FF (µg/in.)	SFC (%)	Yarn strength (cN/tex)
1	28.7	1.09	4.4	13.8	14.5
2	28.5	1.15	3.5	11.9	13.8
3	28.7	1.10	3.7	18.4	13.1
4	30.8	1.13	4.3	9.8	15.2
5	26.5	1.09	3.8	8.4	14.4
6	27.5	1.07	4.5	8.4	14.8
7	29.2	0.98	4.5	16.6	13.2
8	29.0	1.05	4.2	10.9	14.0
9	30.3	1.10	4.4	8.7	14.7
10	28.1	1.01	3.8	15.5	13.4
11	30.6	1.07	4.7	9.0	15.3
12	28.7	1.05	3.9	11.8	14.0
13	28.3	0.97	3.8	13.1	15.1
14	29.0	1.06	3.1	11.3	14.3
15	27.7	1.05	4.7	11.7	14.2
16	29.1	1.05	4.0	11.2	14.9
17	28.6	1.04	4.2	10.8	14.3
18	28.8	1.05	4.1	9.2	17.0
19	28.1	1.03	4.5	7.2	15.3
20	29.0	1.04	5.0	6.8	15.3
21	27.4	0.96	4.6	9.2	14.4
22	27.2	0.98	4.6	10.3	13.1
23	31.7	1.03	3.7	8.9	16.2
24	29.3	1.03	4.4	8.1	16.0
25	29.1	1.05	4.6	5.6	15.8
26	30.8	1.01	3.7	6.8	16.2
27	26.7	1.04	4.8	7.5	15.3
28	30.2	1.06	4.3	5.6	16.7
29	28.7	1.02	3.8	8.7	16.0

Continued

TABLE 7.13 Fiber properties and corresponding yarn strength—cont'd

Sl. no.	FS (cN/tex)	UHML (in.)	FF (μg/in.)	SFC (%)	Yarn strength (cN/tex)
30	29.5	1.02	4.8	7.2	15.4
31	27.5	1.01	4.5	9.7	14.6
32	28.9	1.07	4.6	7.7	15.2
33	28.4	1.02	4.3	10.2	14.1
34	30.3	1.10	4.6	8.4	15.7
35	34.0	1.20	3.8	6.8	18.0
36	26.8	1.00	4.9	6.8	13.8

7.3.1.1 FES modeling

Four parameters of cotton fibers, namely FS, UHML, FF, and SFC have been used as the input parameters to the FES. Three linguistic fuzzy sets namely low, medium, and high were chosen for each of the input parameters in such a way that they are equally spaced and cover the whole input spaces. The triangular membership functions were used for inputs and output. The selection of triangular membership function is attributable to its simplicity as it is merely a collection of three points forming a triangle. Figs. 7.7–7.10 depict the triangular membership curves for FS, UHML, FF, and SFC, respectively. Nine output fuzzy sets (levels 1–9) were considered for yarn strength, which is the output of FES. Fig. 7.11 shows the triangular membership curve for yarn strength. Theoretically, there could be $3^4 = 81$ fuzzy rules, as there are four input variables and each one of them has three linguistic levels. However, to simplify the expert system only 36 fuzzy rules were developed as shown in Table 7.14. As an example, according to the first rule,

IF "FS is low" AND "UHML is low" AND "FF is high" AND "SFC is high," THEN "yarn strength is level 1."

A schematic representation of FES for yarn strength model is displayed in Fig. 7.12. All 36 rules were evaluated simultaneously. The fuzzy sets that represent the outputs of each rule were combined into a single fuzzy set by the process of aggregation. The input of the aggregation process was the list of truncated output functions evaluated by each rule. The "max" function was used to aggregate the output of each rule into a single fuzzy set of the output variable. The aggregate fuzzified output was then converted into a single crisp value by the process of defuzzification. The defuzzification was done by the centroid method, which returns the center of area under the curve to yield the crisp output of yarn strength.

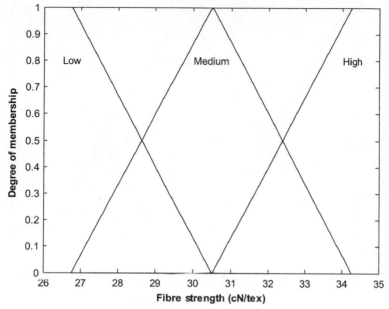

FIG. 7.7   FES membership function for FS.

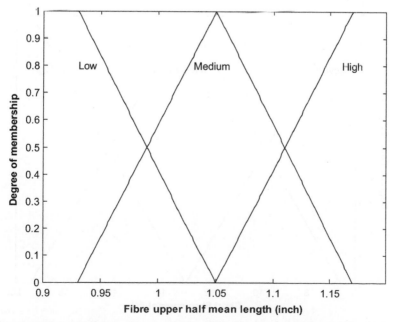

FIG. 7.8   FES membership function for UHML.

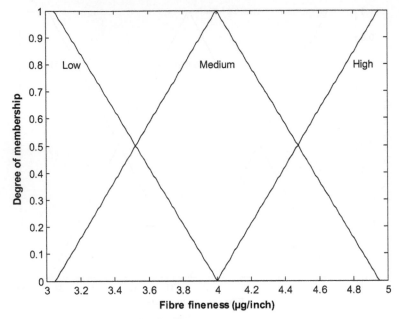

FIG. 7.9 FES membership function for FF.

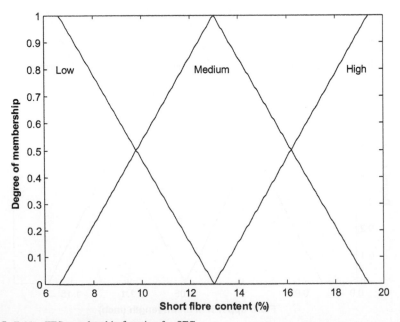

FIG. 7.10 FES membership function for SFC.

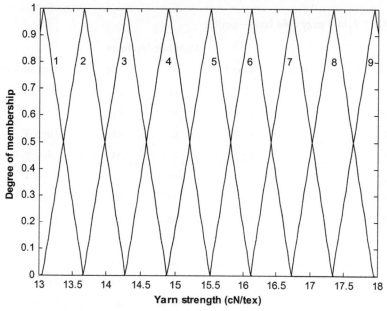

FIG. 7.11 FES membership function for yarn strength.

TABLE 7.14 Fuzzy rule base.

Rule no.	Membership level				
	FS	UHML	FF	SFC	Yarn strength
1	L	L	H	H	Level 1
2	H	H	L	L	Level 9
3	H	M	L	L	Level 7
4	H	L	L	L	Level 4
5	M	H	L	L	Level 6
6	L	H	L	L	Level 3
7	H	H	M	L	Level 7
8	H	H	H	L	Level 4
9	H	H	L	M	Level 8
10	H	H	L	H	Level 6
11	M	L	H	H	Level 3
12	H	L	H	H	Level 5
13	L	M	H	H	Level 2

Continued

TABLE 7.14 Fuzzy rule base—cont'd

Rule no.	Membership level				
	FS	*UHML*	*FF*	*SFC*	*Yarn strength*
14	L	H	H	H	Level 3
15	L	L	M	H	Level 3
16	L	L	L	H	Level 5
17	L	L	H	M	Level 2
18	L	L	H	L	Level 4
19	M	M	M	M	Level 6
20	L	M	M	M	Level 4
21	H	M	M	M	Level 8
22	M	L	M	M	Level 5
23	M	H	M	M	Level 7
24	M	M	L	M	Level 7
25	M	M	H	M	Level 4
26	M	M	M	L	Level 7
27	M	M	M	H	Level 5
28	H	H	M	M	Level 8
29	H	M	L	M	Level 8
30	H	M	M	L	Level 8
31	L	M	M	H	Level 2
32	L	L	M	M	Level 2
33	L	M	H	M	Level 2
34	M	H	L	M	Level 7
35	M	H	M	L	Level 7
36	M	M	L	L	Level 7

7.3.1.2 GFES modeling

A binary-coded GA was used to improve the performance of FES by tuning the membership function distributions of the input and output variables with the help of 30 and 6 randomly partitioned datasets for training and testing, respectively. The same linguistic fuzzy sets and rule base as used in FES were considered for GFES.

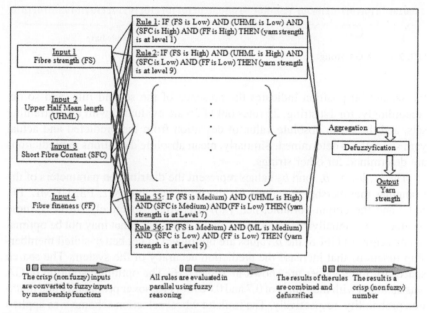

FIG. 7.12 Schematic representation of FES for yarn strength modeling.

An initial population of the binary-coded GA (size $N = 100$) is created at random as shown in Table 7.15. For each individual of the population, starting from the leftmost bit position, eight bits are assigned to indicate each of the b values (i.e., b_1, b_2, b_3, b_4, and b_5) and the next 36 bits represent the rule base. The 1st individual or string of the initial population is shown in Fig. 7.13. The 1 and 0 of the last 36 bits of a GA string denote whether the particular rule from the rule base will be fired or not. In the above example, starting from the left, 0 in the first bit position of the rule base indicates the absence of the first rule, 1 in

TABLE 7.15 A Population of GA strings for GFES.

Sl. no.	GA strings
1	11001011110110110110110111010101101010 01101101110110110111011010010101101 0
2	1101101101101100101111011010 01101101110110100101011011101101110101010101100 0
⋮	⋮
N	01011110110110110110110111010101101010 011011010111011010001010110111101101

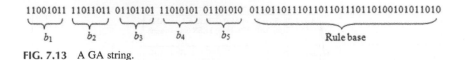

FIG. 7.13 A GA string.

the second bit position indicates the presence of the second rule, and so on. Accordingly, for 1st string, 22 rules out of 36 are evaluated with all 30 training sets, and the mean absolute value of deviation from the predicted and actual yarn strengths is determined. Similarly, mean absolute deviations in predictions are determined for other strings.

The b_1, b_2, b_3, b_4, and b_5 values represent the distribution parameters of the triangular membership functions for the input variables viz. FS, UHML, FF, SFC, and the output variable, i.e., yarn strength, respectively. In FES, these b values are manually designed based on the experience that may not be optimal in any sense. In GFES, the b values are optimized to get better-suited membership functions that increase the prediction accuracy of the system. The search spaces for the b values are shown in Table 7.16. The optimum solution of GA was obtained with the values of 0.7 and 0.001 for crossover probability and mutation probability, respectively. The roulette wheel selection scheme was applied for reproduction operation, and a uniform crossover method was applied.

A MATLAB-based coding was used to execute both FES and GFES models of yarn strength. Figs. 7.14–7.17 show the optimized membership functions for four input parameters and Fig. 7.18 depicts the 9 levels of membership function for yarn strength obtained by the GFES model. Table 7.16 shows the manually designed as well as the optimized b values for FES and GFES models, respectively.

The prediction accuracies of the FES and GFES models were evaluated by calculating the coefficient of determination (R^2) and mean absolute error (%) from the actual and predicted yarn strength values. Table 7.17 shows the

TABLE 7.16 The b values of FES and GFES models.

	FES model	GFES model	
b values	Set value	Search space	Optimum value
b_1	3.75	3.75 ± 1.5	2.833
b_2	0.12	0.12 ± 0.07	0.091
b_3	0.95	0.95 ± 0.25	1.200
b_4	6.40	6.40 ± 4.0	9.200
b_5	0.61	0.61 ± 0.25	0.657

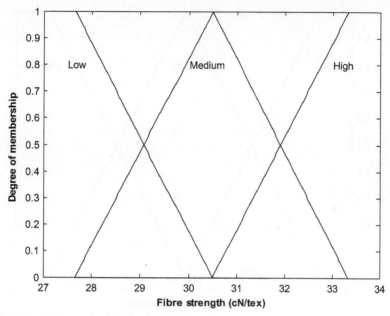

FIG. 7.14   GFES membership function for FS.

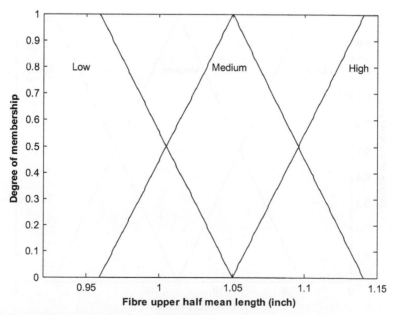

FIG. 7.15   GFES membership function for UHML.

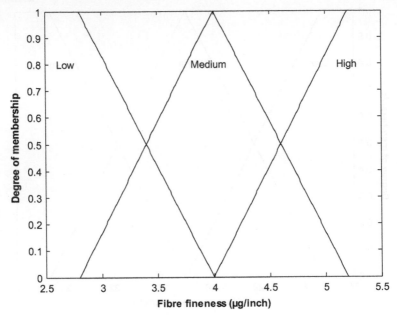

FIG. 7.16 GFES membership function for FF.

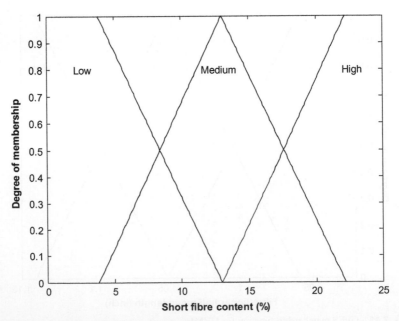

FIG. 7.17 GFES membership function for SFC.

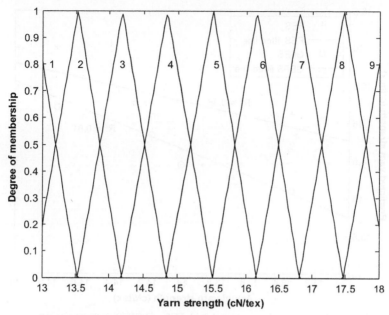

FIG. 7.18 GFES membership function for yarn strength.

TABLE 7.17 Prediction of yarn strength (cN/tex) by FES and GFES models.

Actual yarn strength	FES model		GFES model	
	Predicted strength	*Absolute error (%)*	*Predicted strength*	*Absolute error (%)*
14.0	15.41	10.07	14.45	3.21
14.2	14.82	4.37	14.85	4.58
13.4	15.12	12.84	14.10	5.22
14.3	15.29	6.92	15.27	6.78
14.8	16.01	8.18	15.22	2.84
14.0	15.35	9.64	14.52	3.71
Mean error (%)		8.67		4.39

prediction error of the testing dataset obtained by both the models. The GFES model shows a lower mean error (4.39%) as compared to that of the FES model (8.67%). The prediction results are also depicted in Fig. 7.19 from which it is observed that R^2 is 0.81 for the GFES model, which means that it can explain up to 81% of the total variability of yarn strength. Therefore, the GEFS model of

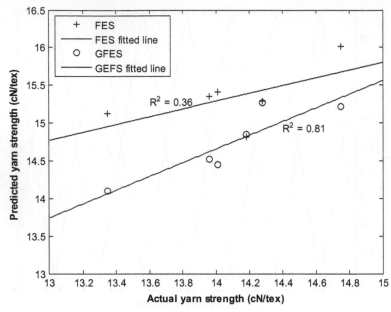

FIG. 7.19   Actual vs. predicted yarn strength with FES and GFES on the testing data.

yarn strength demonstrates a reasonably good degree of prediction consistency. However, a low value of R^2 for the FES model (0.36) signifies a poor prediction consistency.

7.4   MATLAB coding

7.4.1   MATLAB coding of ANFIS application given in Section 7.2.2

```
clc
close all
clear
rotor_data=[27.69    4.4 13.8    10.2    10.3
            27.69    4.4 13.8    21.6    12.03
            27.69    4.4 13.8    29.8    13.77
            29.21    3.5 11.9    10.1    10.1
            29.21    3.5 11.9    21.8    11.94
            29.21    3.5 11.9    29.9    14.09
            27.94    3.7 18.4    10      10.4
            27.94    3.7 18.4    21.7    11.9
            27.94    3.7 18.4    30.3    13.78
            28.7     4.3 9.8     10      9.95
            28.7     4.3 9.8     21.9    11.94
            28.7     4.3 9.8     30      13.85
```

27.69	3.8	8.4	10	10.11
27.69	3.8	8.4	21.7	11.75
27.69	3.8	8.4	29.7	13.66
27.18	4.5	8.4	10.1	10.51
27.18	4.5	8.4	21.8	12.34
27.18	4.5	8.4	29.6	14.38
24.89	4.5	16.6	9.9	11.12
24.89	4.5	16.6	21.8	12.47
24.89	4.5	16.6	29.8	14.44
26.67	4.2	10.9	10	10.16
26.67	4.2	10.9	21.5	12.07
26.67	4.2	10.9	30.1	14.06
27.94	4.4	8.7	10.1	10.5
27.94	4.4	8.7	21.5	12.38
27.94	4.4	8.7	30.2	14.36
25.65	3.8	15.5	9.9	10.78
25.65	3.8	15.5	21.7	12.52
25.65	3.8	15.5	29.6	14.34
27.18	4.7	9	10.1	10.32
27.18	4.7	9	21.9	12.22
27.18	4.7	9	29.9	14.45
26.67	3.9	11.8	10	10.71
26.67	3.9	11.8	21.9	12.45
26.67	3.9	11.8	30	14.31
24.64	3.8	13.1	10.1	10.34
24.64	3.8	13.1	21.7	11.9
24.64	3.8	13.1	30.5	13.78
26.92	3.1	11.3	9.9	10.3
26.92	3.1	11.3	21.8	11.78
26.92	3.1	11.3	29.8	13.78
26.67	4.7	11.7	10.2	10.29
26.67	4.7	11.7	21.7	12.18
26.67	4.7	11.7	30.4	14.31
26.67	4	11.2	10	10.42
26.67	4	11.2	21.8	12.05
26.67	4	11.2	30	14.11
26.42	4.2	10.8	10	10.66
26.42	4.2	10.8	21.8	12.23
26.42	4.2	10.8	30	14.21
26.67	4.1	9.2	10	10.22
26.67	4.1	9.2	21.9	12.06
26.67	4.1	9.2	29.6	14.16
26.16	4.5	7.2	10.2	10.24

26.16	4.5	7.2	22.3	12.07
26.16	4.5	7.2	30.3	13.89
26.42	5	6.8	10.1	10.27
26.42	5	6.8	22	12.11
26.42	5	6.8	30.2	13.83
24.38	4.6	9.2	10.1	10.7
24.38	4.6	9.2	22.1	12.34
24.38	4.6	9.2	30.2	13.97
24.89	4.6	10.3	10.1	11.2
24.89	4.6	10.3	22.2	12.69
24.89	4.6	10.3	29.8	14.35
26.16	3.7	8.9	10	10.26
26.16	3.7	8.9	22.6	11.66
26.16	3.7	8.9	30.4	13.55
26.16	4.4	8.1	10.1	10.04
26.16	4.4	8.1	22	11.78
26.16	4.4	8.1	30.1	13.49
26.67	4.6	5.6	10.1	10.1
26.67	4.6	5.6	22.2	12.1
26.67	4.6	5.6	30.2	13.92
25.65	3.7	6.8	10	10.18
25.65	3.7	6.8	21.9	11.88
25.65	3.7	6.8	29.9	13.46
26.42	4.8	7.5	10	10.61
26.42	4.8	7.5	22	12.33
26.42	4.8	7.5	30	14.19
26.92	4.3	5.6	10.2	10.1
26.92	4.3	5.6	21.8	11.98
26.92	4.3	5.6	30.4	14.04
25.91	3.8	8.7	10.1	10.02
25.91	3.8	8.7	22.2	11.77
25.91	3.8	8.7	29.9	13.54
25.91	4.8	7.2	10.1	10.4
25.91	4.8	7.2	22	12.18
25.91	4.8	7.2	30.6	14.22
25.65	4.5	9.7	10	10.58
25.65	4.5	9.7	22.1	12.16
25.65	4.5	9.7	30.2	14.42
27.18	4.6	7.7	10	10.07
27.18	4.6	7.7	22.4	11.97
27.18	4.6	7.7	30.5	13.94
25.91	4.3	10.2	10.1	10.54
25.91	4.3	10.2	21.9	11.97

```
                25.91    4.3 10.2   29.7   13.68
                27.94    4.6 8.4    10      9.94
                27.94    4.6 8.4    22     11.66
                27.94    4.6 8.4    30.6   13.76
                30.48    3.8 6.8    10      9.78
                30.48    3.8 6.8    22.1   11.34
                30.48    3.8 6.8    30.4   13.23
                25.4     4.9 6.8    10.1   10.62
                25.4     4.9 6.8    22.2   12.38
                25.4     4.9 6.8    30.1   14.21];
[s1,s2]=size(rotor_data);
input_array=rotor_data(:,1:4);
target_array=rotor_data(:,end);
n=s1;
k=12;
p=0;
q=0;
r=0;
s=1:n;
for i=1:n
    if rem(i,k)==0
        p=p+k;
        q=q+1;
        tst(q)=s(p);
    else
        r=r+1;
        trn(r)=s(i);
    end
end
input_array_training=input_array(trn,:);
target_array_training=target_array(trn);
input_array_testing=input_array(tst,:);
target_array_testing=target_array(tst);
trn_data=[input_array_training target_array_training];
input_array_testing;
target_array_testing;
opt = genfisOptions('GridPartition');
opt.NumMembershipFunctions = [2 2 2 2];
opt.InputMembershipFunctionType = ["gbellmf" "gbellmf" "gbellmf"
"gbellmf"];
in_fis=genfis(input_array_training,target_array_training,opt);
in_fis.input
trnOpt=[10000 1e-3 0.01 0.9 1.1];
dispOpt=[1 1 1 1];
```

```
[fis,error,stepsize]=anfis(trn_data,in_fis,trnOpt,dispOpt,
[],1);
y=target_array_testing;
y1=evalfis(input_array_testing,fis);
z=target_array_training;
z1=evalfis(input_array_training,fis);
Prediction=[y y1 (y-y1)]
mean_accuracy_testing=mean(100-((abs(y-y1)./y)*100))
mean_accuracy_training=mean(100-((abs(z-z1)./z)*100))
figure(1)
[x1,mf] = plotmf(in_fis,'input',1);
subplot(2,1,1)
plot(x1,mf,'k')
xlabel('Upper half mean length(mm)','FontSize',10,'FontWeight','
bold')
ylabel('Membership function','FontSize',10,'FontWeight','bold')
[xf1,mf] = plotmf(fis,'input',1);
subplot(2,1,2)
plot(xf1,mf,'k')
xlabel('Upper half mean length(mm)','FontSize',10,'FontWeight',
'bold')
ylabel('Membership function','FontSize',10,'FontWeight','bold')
set(gcf,'color','w')
figure(2)
[x2,mf] = plotmf(in_fis,'input',2);
subplot(2,1,1)
plot(x2,mf,'k')
xlabel('Micronaire','FontSize',10,'FontWeight','bold')
ylabel('Membership function','FontSize',10,'FontWeight','bold')
[xf2,mf] = plotmf(fis,'input',2);
subplot(2,1,2)
plot(xf2,mf,'k')
xlabel('Micronaire','FontSize',10,'FontWeight','bold')
ylabel('Membership function','FontSize',10,'FontWeight','bold')
set(gcf,'color','w')
figure(3)
[x3,mf] = plotmf(in_fis,'input',3);
subplot(2,1,1)
plot(x3,mf,'k')
xlabel('Short fibre(%)','FontSize',10,'FontWeight','bold')
ylabel('Membership function','FontSize',10,'FontWeight','bold')
[xf3,mf] = plotmf(fis,'input',3);
subplot(2,1,2)
```

```
plot(xf3,mf,'k')
xlabel('Short fibre(%)','FontSize',10,'FontWeight','bold')
ylabel('Membership function','FontSize',10,'FontWeight','bold')
set(gcf,'color','w')
figure(4)
[x4,mf] = plotmf(in_fis,'input',4);
subplot(2,1,1)
plot(x4,mf,'k')
xlabel('Yarn English count','FontSize',10,'FontWeight','bold')
ylabel('Membership function','FontSize',10,'FontWeight','bold')
[xf4,mf] = plotmf(fis,'input',4);
subplot(2,1,2)
plot(xf4,mf,'k')
xlabel('Yarn English count','FontSize',10,'FontWeight','bold')
ylabel('Membership function','FontSize',10,'FontWeight','bold')
set(gcf,'color','w')
```

7.4.2   MATLAB coding of GFES application given in Section 7.3.1.2

```
clc
close all
clear
format short g
dataset=[28.7    1.09    4.4    13.8    14.5
         28.5    1.15    3.5    11.9    13.8
         28.7    1.1     3.7    18.4    13.1
         30.8    1.13    4.3     9.8    15.2
         26.5    1.09    3.8     8.4    14.4
         27.5    1.07    4.5     8.4    14.8
         29.2    0.98    4.5    16.6    13.2
         29      1.05    4.2    10.9    14
         30.3    1.1     4.4     8.7    14.7
         28.1    1.01    3.8    15.5    13.4
         30.6    1.07    4.7     9      15.3
         28.7    1.05    3.9    11.8    14
         28.3    0.97    3.8    13.1    15.1
         29      1.06    3.1    11.3    14.3
         27.7    1.05    4.7    11.7    14.2
         29.1    1.05    4      11.2    14.9
         28.6    1.04    4.2    10.8    14.3
         28.8    1.05    4.1     9.2    17
         28.1    1.03    4.5     7.2    15.3
         29      1.04    5       6.8    15.3
         27.4    0.96    4.6     9.2    14.4
```

```
                27.2    0.98   4.6    10.3   13.1
                31.7    1.03   3.7    8.9    16.2
                29.3    1.03   4.4    8.1    16
                29.1    1.05   4.6    5.6    15.8
                30.8    1.01   3.7    6.8    16.2
                26.7    1.04   4.8    7.5    15.3
                30.2    1.06   4.3    5.6    16.7
                28.7    1.02   3.8    8.7    16
                29.5    1.02   4.8    7.2    15.4
                27.5    1.01   4.5    9.7    14.6
                28.9    1.07   4.6    7.7    15.2
                28.4    1.02   4.3    10.2   14.1
                30.3    1.1    4.6    8.4    15.7
                34      1.2    3.8    6.8    18
                26.8    1      4.9    6.8    13.8];
[s1,s2]=size(dataset);
random_sample=randperm(s1);%Data randomisation
dataset=dataset(random_sample,:);
input_data_trn=dataset(1:30, 1:4);
output_data_trn=dataset(1:30, end);
input_data_tst=dataset(31:36, 1:4);
output_data_tst=dataset(31:36, end);
%Initialization of Parameters
pop_size=100; %Population size
bit_size=[8 8 8 8 8 36]; % Bits size
string_length=sum(bit_size); %length of strings
ub=[5.25 0.19 1.2 10.4 0.86];%upper bound values
lb=[2.25 0.05 0.7 2.4 0.36];%lower bound values
pm=0.001;%Mutation probability
pc=0.7;    %Crossover probability
gen=1      %Genaration number initial value
maxgen=100;%Maximum no. of generation
pop=round(rand(pop_size,string_length));%Random population
% Decoding of Old Population
while maxgen >= gen
dec_val(pop_size,length(bit_size))=0;
dec_val(:,:)=0;
m=1;
for j=1:length(bit_size)
    cg=m;
    for i=bit_size(j)-1:-1:0
     dec_val(:,j)=dec_val(:,j)+ 2^i.* pop(:,cg);
     cg=cg+1;
```

```
        m=cg;
    end
end
dec_val;% decoded value of substrings
%Calculation of parameter values
for i=1:length(ub)
    accuracy(i) = (ub(i)-lb(i))/((2^bit_size(i))-1);
    X(i,:) =lb(i)+ dec_val(:,i).* accuracy(i);
end
X_values=X';
for i=1:pop_size
%FIS Starts Here
%Connecting to fuzzy toolbox (FIS)
a=newfis('strength');
%Add Fibre strength (Input 1) to FIS
d1=X(1,i);
x1_min=min(input_data_trn(:,1));
x1_max=max(input_data_trn(:,1));
x1_mid=(x1_min+x1_max)/2;
x1_range=[x1_mid-d1 x1_mid+d1];
x1_mf1=[x1_mid-2*d1    x1_mid-d1    x1_mid];
x1_mf2=[x1_mid-d1       x1_mid    x1_mid+d1];
x1_mf3=[x1_mid         x1_mid+d1 x1_mid+2*d1];
a=addvar(a,'input','Fibre strength',x1_range);
a=addmf(a,'input',1,'low','trimf',x1_mf1);
a=addmf(a,'input',1,'medium','trimf',x1_mf2);
a=addmf(a,'input',1,'high','trimf',x1_mf3);
%Add Upper Half Mean length (Input 2) to FIS
d2=X(2,i);
x2_min=min(input_data_trn(:,2));
x2_max=max(input_data_trn(:,2));
x2_mid=(x2_min+x2_max)/2;
x2_range=[x2_mid-d2 x2_mid+d2];
x2_mf1=[x2_mid-2*d2    x2_mid-d2    x2_mid];
x2_mf2=[x2_mid-d2       x2_mid    x2_mid+d2];
x2_mf3=[x2_mid         x2_mid+d2 x2_mid+2*d2];
a=addvar(a,'input','UHML',x2_range);
a=addmf(a,'input',2,'low','trimf',x2_mf1);
a=addmf(a,'input',2,'medium','trimf',x2_mf2);
a=addmf(a,'input',2,'high','trimf',x2_mf3);
%Add Mic (Input 3) to FIS
d3=X(3,i);
x3_min=min(input_data_trn(:,3));
```

```
x3_max=max(input_data_trn(:,3));
x3_mid=(x3_min+x3_max)/2;
x3_range=[x3_mid-d3 x3_mid+d3];
x3_mf1=[x3_mid-2*d3    x3_mid-d3    x3_mid];
x3_mf2=[x3_mid-d3       x3_mid      x3_mid+d3];
x3_mf3=[x3_mid          x3_mid+d3   x3_mid+2*d3];
a=addvar(a,'input','Mic',x3_range);
a=addmf(a,'input',3,'low','trimf',x3_mf1);
a=addmf(a,'input',3,'medium','trimf',x3_mf2);
a=addmf(a,'input',3,'high','trimf',x3_mf3);
%Add SFC (Input 4) to FIS
d4=X(4,i);
x4_min=min(input_data_trn(:,4));
x4_max=max(input_data_trn(:,4));
x4_mid=(x4_min+x4_max)/2;
x4_range=[x4_mid-d4 x4_mid+d4];
x4_mf1=[x4_mid-2*d4    x4_mid-d4    x4_mid];
x4_mf2=[x4_mid-d4       x4_mid      x4_mid+d4];
x4_mf3=[x4_mid          x4_mid+d4   x4_mid+2*d4];
a=addvar(a,'input','SFC',x4_range);
a=addmf(a,'input',4,'low','trimf',x4_mf1);
a=addmf(a,'input',4,'medium','trimf',x4_mf2);
a=addmf(a,'input',4,'high','trimf',x4_mf3);
%Add Yarn-tenacity (Output) to FIS
d=X(5,i);
y_min=min(output_data_trn);
y_max=max(output_data_trn);
y_mid=(y_min+y_max)/2;
y_range=[y_mid-d y_mid+d];
y_mf1=[y_mid-5*d    y_mid-4*d    y_mid-3*d];
y_mf2=[y_mid-4*d    y_mid-3*d    y_mid-2*d];
y_mf3=[y_mid-3*d    y_mid-2*d    y_mid-d];
y_mf4=[y_mid-2*d    y_mid-d      y_mid];
y_mf5=[y_mid-d      y_mid        y_mid+d];
y_mf6=[y_mid        y_mid+d      y_mid+2*d];
y_mf7=[y_mid+d      y_mid+2*d    y_mid+3*d];
y_mf8=[y_mid+2*d    y_mid+3*d    y_mid+4*d];
y_mf9=[y_mid+3*d    y_mid+4*d    y_mid+5*d];
a=addvar(a,'output','yarn strength',y_range);
a=addmf(a,'output',1,'1','trimf',y_mf1);
a=addmf(a,'output',1,'2','trimf',y_mf2);
a=addmf(a,'output',1,'3','trimf',y_mf3);
a=addmf(a,'output',1,'4','trimf',y_mf4);
```

```
a=addmf(a,'output',1,'5','trimf',y_mf5);
a=addmf(a,'output',1,'6','trimf',y_mf6);
a=addmf(a,'output',1,'7','trimf',y_mf7);
a=addmf(a,'output',1,'8','trimf',y_mf8);
a=addmf(a,'output',1,'9','trimf',y_mf9);
rule_list=[1 1 3 3 1 1 1
           3 3 1 1 9 1 1
           3 2 1 1 7 1 1
           3 1 1 1 4 1 1
           2 3 1 1 6 1 1
           1 3 1 1 3 1 1
           3 3 2 1 7 1 1
           3 3 3 1 4 1 1
           3 3 1 2 8 1 1
           3 3 1 3 6 1 1
           2 1 3 3 3 1 1
           3 1 3 3 5 1 1
           1 2 3 3 2 1 1
           1 3 3 3 3 1 1
           1 1 2 3 3 1 1
           1 1 1 3 5 1 1
           1 1 3 2 2 1 1
           1 1 3 1 4 1 1
           2 2 2 2 6 1 1
           1 2 2 2 4 1 1
           3 2 2 2 8 1 1
           2 1 2 2 5 1 1
           2 3 2 2 7 1 1
           2 2 1 2 7 1 1
           2 2 3 2 4 1 1
           2 2 2 1 7 1 1
           2 2 2 3 5 1 1
           3 3 2 2 8 1 1
           3 2 1 2 8 1 1
           3 2 2 1 8 1 1
           1 2 2 3 2 1 1
           1 1 2 2 2 1 1
           1 2 3 2 2 1 1
           2 3 1 2 7 1 1
           2 3 2 1 7 1 1
           2 2 1 1 7 1 1];
%rule selection
mm= sum(bit_size(1:5))+1;
```

```
cq=1;
clear sel_list;
for rr=1:length(rule_list)
    if pop(i,mm)==1
        sel_list(cq,:)=rule_list(rr,:);
        cq=cq+1;
    end
    mm=mm+1;
end
%Fitnes Evaluation
a=addrule(a,sel_list);
y=output_data_trn;
y1=evalfis(input_data_trn, a);
Prediction=[y y1];
mean_dev=sum(abs(y-y1))/length(y);
fx(i)= mean_dev;
end
Fx=1./(1+fx.^2);
fit_val=[fx' Fx'];% Fitness values
old_ratio=mean(Fx)/max(Fx);% Old Ration
%Cumulative probability
A=Fx./mean(Fx);
B=A./pop_size;%Probability of selection
for i=1:pop_size
C(i)=sum(B(1:i)); %Cumulative probability
end
D=rand(pop_size,1);%Random number for Roulette wheel selection
%Roullette Wheel Selection
for i=1:pop_size % Finding the position of the random no.
    for j=1:pop_size-1
        if D(i)>=C(j) && D(i)<=C(j+1)
            E(i)=j+1;
        elseif D(i)<=C(1)
            E(i)=1;
        end
    end
end
for i=1:pop_size %The count of the String in the mating pool
    counter=0;
    F(i)=counter;
    for j=1:pop_size
        if E(j) ==   i
            counter= counter+1;
```

```
                F(i)=counter;
          end
      end
end
strong_pop=pop(E,:);%Mating Pool with stronger parents
%Cross Over
cross=rand(pop_size/2,1);%Toss for selection of strings
for i=1:length(cross)
    if cross(i) <= pc
          x_over(i,:) = 'Y';
    else
          x_over(i,:) = 'N';
    end
end
x_over;
toss_cross=round(rand(pop_size/2,string_length));
m=1;
for i=1:pop_size/2
    if x_over(i)== 'Y'
        for n=1:string_length
            if toss_cross(i,n)==1   % Tossing for crossover
                temp=strong_pop(m,n);%Swapping
                strong_pop(m,n)=strong_pop(m+1,n);
                strong_pop(m+1,n)=temp;
            end
        end
    end
    m= m+2;
end
corss_pop=strong_pop;
%Mutation
toss_m= rand(pop_size,string_length);
k=1;
m_x(k)=0;
m_y(k)=0;
for i=1:pop_size
    for j=1:string_length
        if toss_m(i,j) <= pm
            m_x(k)=i;
            m_y(k)=j;
            k=k+1;
            if corss_pop(i,j) == 1
                corss_pop(i,j)= 0;
```

```
            elseif corss_pop(i,j) == 0
                corss_pop(i,j) = 1;
            end
        end
    end
end
mutated_bits= [m_x' m_y']; % Mutation points
new_pop = corss_pop;  % Mutated population
%Decoding of new population
new_dec_val(pop_size,length(bit_size))=0;
new_dec_val(:,:)=0;
m=1;
for j=1:length(bit_size)
    cg=m;
    for i=bit_size(j)-1:-1:0
     new_dec_val(:,j)=new_dec_val(:,j)+ 2^i.* new_pop(:,cg);
     cg=cg+1;
     m=cg;
    end
end
new_dec_val;%decoded value of substrings
%Calculation of parameter values
for i=1:length(ub)
    new_X(i,:) =lb(i)+ new_dec_val(:,i).* accuracy(i);
end
new_X_values=new_X';
%Fitness Evaluation
for i=1:pop_size
%FIS Starts Again
a=newfis('strength');
%Add Fibre strength (Input 1) to FIS
d1=new_X(1,i);
x1_min=min(input_data_trn(:,1));
x1_max=max(input_data_trn(:,1));
x1_mid=(x1_min+x1_max)/2;
x1_range=[x1_mid-d1 x1_mid+d1];
x1_mf1=[x1_mid-2*d1   x1_mid-d1    x1_mid];
x1_mf2=[x1_mid-d1     x1_mid     x1_mid+d1];
x1_mf3=[x1_mid        x1_mid+d1  x1_mid+2*d1];
a=addvar(a,'input','Fibre strength',x1_range);
a=addmf(a,'input',1,'low','trimf',x1_mf1);
a=addmf(a,'input',1,'medium','trimf',x1_mf2);
a=addmf(a,'input',1,'high','trimf',x1_mf3);
```

```
%Add Upper Half Mean length (Input 2) to FIS
d2=new_X(2,i);
x2_min=min(input_data_trn(:,2));
x2_max=max(input_data_trn(:,2));
x2_mid=(x2_min+x2_max)/2;
x2_range=[x2_mid-d2 x2_mid+d2];
x2_mf1=[x2_mid-2*d2    x2_mid-d2    x2_mid];
x2_mf2=[x2_mid-d2        x2_mid      x2_mid+d2];
x2_mf3=[x2_mid          x2_mid+d2    x2_mid+2*d2];
a=addvar(a,'input','UHML',x2_range);
a=addmf(a,'input',2,'low','trimf',x2_mf1);
a=addmf(a,'input',2,'medium','trimf',x2_mf2);
a=addmf(a,'input',2,'high','trimf',x2_mf3);
%Add Mic (Input 3) to FIS
d3=new_X(3,i);
x3_min=min(input_data_trn(:,3));
x3_max=max(input_data_trn(:,3));
x3_mid=(x3_min+x3_max)/2;
x3_range=[x3_mid-d3 x3_mid+d3];
x3_mf1=[x3_mid-2*d3    x3_mid-d3    x3_mid];
x3_mf2=[x3_mid-d3        x3_mid      x3_mid+d3];
x3_mf3=[x3_mid          x3_mid+d3    x3_mid+2*d3];
a=addvar(a,'input','Mic',x3_range);
a=addmf(a,'input',3,'low','trimf',x3_mf1);
a=addmf(a,'input',3,'medium','trimf',x3_mf2);
a=addmf(a,'input',3,'high','trimf',x3_mf3);
%Add SFC(Input 4) to FIS
d4=new_X(4,i);
x4_min=min(input_data_trn(:,4));
x4_max=max(input_data_trn(:,4));
x4_mid=(x4_min+x4_max)/2;
x4_range=[x4_mid-d4 x4_mid+d4];
x4_mf1=[x4_mid-2*d4    x4_mid-d4    x4_mid];
x4_mf2=[x4_mid-d4        x4_mid      x4_mid+d4];
x4_mf3=[x4_mid          x4_mid+d4    x4_mid+2*d4];
a=addvar(a,'input','SFC',x4_range);
a=addmf(a,'input',4,'low','trimf',x4_mf1);
a=addmf(a,'input',4,'medium','trimf',x4_mf2);
a=addmf(a,'input',4,'high','trimf',x4_mf3);
%Add Yarn-tenacity (Output) to FIS
d=new_X(5,i);
y_min=min(output_data_trn);
y_max=max(output_data_trn);
```

```
y_mid=(y_min+y_max)/2;
y_range=[y_min y_max];
y_mf1=[y_mid-5*d    y_mid-4*d    y_mid-3*d];
y_mf2=[y_mid-4*d    y_mid-3*d    y_mid-2*d];
y_mf3=[y_mid-3*d    y_mid-2*d    y_mid-d];
y_mf4=[y_mid-2*d    y_mid-d      y_mid];
y_mf5=[y_mid-d      y_mid        y_mid+d];
y_mf6=[y_mid        y_mid+d      y_mid+2*d];
y_mf7=[y_mid+d      y_mid+2*d    y_mid+3*d];
y_mf8=[y_mid+2*d    y_mid+3*d    y_mid+4*d];
y_mf9=[y_mid+3*d    y_mid+4*d    y_mid+5*d];
a=addvar(a,'output','yarn strength',y_range);
a=addmf(a,'output',1,'1','trimf',y_mf1);
a=addmf(a,'output',1,'2','trimf',y_mf2);
a=addmf(a,'output',1,'3','trimf',y_mf3);
a=addmf(a,'output',1,'4','trimf',y_mf4);
a=addmf(a,'output',1,'5','trimf',y_mf5);
a=addmf(a,'output',1,'6','trimf',y_mf6);
a=addmf(a,'output',1,'7','trimf',y_mf7);
a=addmf(a,'output',1,'8','trimf',y_mf8);
a=addmf(a,'output',1,'9','trimf',y_mf9);
% rule selection
mm= sum(bit_size(1:5))+1;
cq=1;
clear sel_list;
for rr=1:length(rule_list)
    if new_pop(i,mm)==1
        sel_list(cq,:)=rule_list(rr,:);
        cq=cq+1;
    end
    mm=mm+1;
end
%Fitnes Evaluation
a=addrule(a,sel_list);
rar=1;
y=output_data_trn;
y1=evalfis(input_data_trn, a);
Prediction=[y y1];
mean_dev=sum(abs(y-y1))/length(y);
new_fx(i)= mean_dev;
end
new_Fx=1./(1+new_fx.^2);
new_fit_val=[new_fx,new_Fx];% Fitness values of new population
```

```
new_ratio=mean(new_Fx)/max(new_Fx)% New Ratio
pop=new_pop;
    if new_ratio>=0.95
        break;
    end
gen=gen+1
end
[fitness,i]=min(new_fx);
fitness;
X_value=new_X(:,i)
% validation
d1=X_value(1);
d2=X_value(2);
d3=X_value(3);
d4=X_value(4);
d-X_value(5);
%Connecting to fuzzy toolbox
a=newfis('Yarn strength');
%Add Fibre strength (Input 1) to FIS
x1_min=min(input_data_trn(:,1));
x1_max=max(input_data_trn(:,1));
x1_mid=(x1_min+x1_max)/2;
x1_range=[x1_mid-d1 x1_mid+d1];
x1_mf1=[x1_mid-2*d1    x1_mid-d1     x1_mid];
x1_mf2=[x1_mid-d1       x1_mid       x1_mid+d1];
x1_mf3=[x1_mid         x1_mid+d1     x1_mid+2*d1];
a=addvar(a,'input','Fibre bundle tenacity (cN/tex)',x1_range);
a=addmf(a,'input',1,'low','trimf',x1_mf1);
a=addmf(a,'input',1,'medium','trimf',x1_mf2);
a=addmf(a,'input',1,'high','trimf',x1_mf3);
%Add Upper Half Mean length (Input 2) to FIS
x2_min=min(input_data_trn(:,2));
x2_max=max(input_data_trn(:,2));
x2_mid=(x2_min+x2_max)/2;
x2_range=[x2_mid-d2 x2_mid+d2];
x2_mf1=[x2_mid-2*d2    x2_mid-d2     x2_mid];
x2_mf2=[x2_mid-d2       x2_mid       x2_mid+d2];
x2_mf3=[x2_mid         x2_mid+d2     x2_mid+2*d2];
a=addvar(a,'input','Upper half mean length (mm)',x2_range);
a=addmf(a,'input',2,'low','trimf',x2_mf1);
a=addmf(a,'input',2,'medium','trimf',x2_mf2);
a=addmf(a,'input',2,'high','trimf',x2_mf3);
%Add Mic (Input 3) to FIS
```

```
x3_min=min(input_data_trn(:,3));
x3_max=max(input_data_trn(:,3));
x3_mid=(x3_min+x3_max)/2;
x3_range=[x3_mid-d3 x3_mid+d3];
x3_mf1=[x3_mid-2*d3    x3_mid-d3    x3_mid];
x3_mf2=[x3_mid-d3       x3_mid       x3_mid+d3];
x3_mf3=[x3_mid          x3_mid+d3    x3_mid+2*d3];
a=addvar(a,'input','Mic',x3_range);
a=addmf(a,'input',3,'low','trimf',x3_mf1);
a=addmf(a,'input',3,'medium','trimf',x3_mf2);
a=addmf(a,'input',3,'high','trimf',x3_mf3);
%Add SFC (Input 4) to FIS
x4_min=min(input_data_trn(:,4));
x4_max=max(input_data_trn(:,4));
x4_mid=(x4_min+x4_max)/2;
x4_range=[x4_mid-d4 x4_mid+d4];
x4_mf1=[x4_mid-2*d4    x4_mid-d4    x4_mid];
x4_mf2=[x4_mid-d4       x4_mid       x4_mid+d4];
x4_mf3=[x4_mid          x4_mid+d4    x4_mid+2*d4];
a=addvar(a,'input','SFC',x4_range);
a=addmf(a,'input',4,'low','trimf',x4_mf1);
a=addmf(a,'input',4,'medium','trimf',x4_mf2);
a=addmf(a,'input',4,'high','trimf',x4_mf3);
%Add Yarn-tenacity (Output) to FIS
y_min=min(output_data_trn);
y_max=max(output_data_trn);
y_mid=(y_min+y_max)/2;
y_range=[y_min y_max];
y_mf1=[y_mid-5*d    y_mid-4*d      y_mid-3*d];
y_mf2=[y_mid-4*d    y_mid-3*d      y_mid-2*d];
y_mf3=[y_mid-3*d    y_mid-2*d      y_mid-d];
y_mf4=[y_mid-2*d    y_mid-d        y_mid];
y_mf5=[y_mid-d      y_mid          y_mid+d];
y_mf6=[y_mid        y_mid+d        y_mid+2*d];
y_mf7=[y_mid+d      y_mid+2*d      y_mid+3*d];
y_mf8=[y_mid+2*d    y_mid+3*d      y_mid+4*d];
y_mf9=[y_mid+3*d    y_mid+4*d      y_mid+5*d];
a=addvar(a,'output','Yarn strength',y_range);
a=addmf(a,'output',1,'1','trimf',y_mf1);
a=addmf(a,'output',1,'2','trimf',y_mf2);
a=addmf(a,'output',1,'3','trimf',y_mf3);
a=addmf(a,'output',1,'4','trimf',y_mf4);
a=addmf(a,'output',1,'5','trimf',y_mf5);
```

```
a=addmf(a,'output',1,'6','trimf',y_mf6);
a=addmf(a,'output',1,'7','trimf',y_mf7);
a=addmf(a,'output',1,'8','trimf',y_mf8);
a=addmf(a,'output',1,'9','trimf',y_mf9);
a=addrule(a, rule_list);
y=output_data_trn;
y1=evalfis(input_data_trn, a);
Prediction_trn=[y y1]
mean_dev_trn=sum(abs(y-y1))/length(y)
error_trn=100*mean(abs(y-y1)./y)
y3=output_data_tst;
y4=evalfis(input_data_tst, a);
Prediction_tst=[y3 y4]
mean_dev_tst=sum(abs(y3-y4))/length(y3)
error_tst=100*mean(abs(y3-y4)./y3)
```

7.5 Summary

The chapter offers an in-depth exploration of two powerful hybrid computational models, viz., ANFIS and GFES. ANFIS combines the strengths of fuzzy logic and ANNs, whereas GFES combines the power of genetic algorithms and fuzzy expert systems. By combining different AI techniques, it is possible to develop systems that capitalize on the advantages of each component. This chapter equips readers with a deep understanding of ANFIS and GFES by providing a comprehensive overview of their principles, architectures, and applications.

References

Banerjee, D., Ghosh, A., & Das, S. (2012). Yarn strength modelling using genetic fuzzy expert system. *Journal of The Institution of Engineers (India): Series E, 93,* 83–90.

Damousis, I. G., Satsios, K. J., Labridis, D. P., & Dokopoulos, P. S. (2002). Combined fuzzy logic and genetic algorithm techniques – Application to an electromagnetic field problem. *Fuzzy Sets and Systems, 129,* 371–386.

Ghosh, A. (2010). Forecasting of cotton yarn properties using intelligent machines. *Research Journal of Textile and Apparel, 14*(3), 55–61.

Jang, J. S. R. (1993). ANFIS: Adaptive-network-based fuzzy inference system. *IEEE Transactions on Systems, Man, and Cybernetics, 23*(3), 665–685.

Jang, J. S. R., Sun, C. T., & Mizutani, E. (1997). *Neuro-fuzzy and soft computing: A computational approach to learning and machine intelligence.* Upper Saddle River, NJ: Prentice Hall.

Shimojima, K., Fukuda, T., & Hasegawa, Y. (1995). Self-tuning fuzzy modelling with adaptive membership function, rules and hierarchical structure based on genetic algorithm. *Fuzzy Sets and Systems, 71*(3), 295–309.

Index

Note: Page numbers followed by *f* indicate figures and *t* indicate tables.

Printed and bound by CPI Group (UK) Ltd, Croydon, CR0 4YY

03/10/2024

01040847-0004